Exploring Bioactive Polyphenolic Compounds in Food and Natural Real-World Samples II

Guest Editors

Francesco Cacciola
Katia Arena

MDPI

Basel • Beijing • Wuhan • Barcelona • Belgrade • Novi Sad • Cluj • Manchester

Guest Editors
Francesco Cacciola
Department of Chemical,
Biological, Pharmaceutical
and Environmental Sciences
University of Messina
Messina
Italy

Katia Arena
Department of Chemical,
Biological, Pharmaceutical
and Environmental Sciences
University of Messina
Messina
Italy

Editorial Office
MDPI AG
Grosspeteranlage 5
4052 Basel, Switzerland

This is a reprint of the Special Issue, published open access by the journal *Molecules* (ISSN 1420-3049), freely accessible at: https://www.mdpi.com/journal/molecules/special_issues/74D33WSL60.

For citation purposes, cite each article independently as indicated on the article page online and as indicated below:

Lastname, A.A.; Lastname, B.B. Article Title. *Journal Name* **Year**, *Volume Number*, Page Range.

ISBN 978-3-7258-8153-6 (Hbk)
ISBN 978-3-7258-8154-3 (PDF)
https://doi.org/10.3390/books978-3-7258-8154-3

Contents

About the Editors

Francesco Cacciola

Francesco Cacciola is Full Professor of Food Chemistry at the "Dipartimento di Scienze Chimiche, Biologiche, Farmaceutiche ed Ambientali" of the University of Messina, Italy. He received his Ph.D. in Food and Safety Chemistry at the University of Messina in 2009. In September 2018, The International Scientific journal *The Analytical Scientist* included him in the international list of the 40 most influential scientists "under 40" in *The Analytical Scientist's* 2018 Power List "Top 40 under 40" for the application of advanced analytical techniques to the characterization of food and natural products. His research interests include conventional and innovative LC techniques for the study of bioactive compounds, e.g., polyphenols contained in complex natural matrices.

Katia Arena

Katia Arena is Assistant Professor of Food Chemistry at the "Dipartimento di Scienze Chimiche, Biologiche, Farmaceutiche ed Ambientali" of the University of Messina, Italy. She received her master's degree in Chemistry and Pharmaceutical Technologies at the University of Messina in 2016. In 2021, she earned her PhD in Chemical Sciences, defending a thesis entitled "Foodomics: LC×LC Approach in Modern Food Science". Her research activity is focused on the implementation of comprehensive two-dimensional liquid chromatography coupled with mass spectrometry for the determination of the composition of complex food and natural matrices.

molecules

MDPI

Editorial

Exploring Bioactive Polyphenolic Compounds in Food and Natural Real-World Samples II: Molecular Diversity, Functionality, and Future Directions

Francesco Cacciola * and Katia Arena

Department of Chemical, Biological, Pharmaceutical and Environmental Sciences, Messina Institute of Technology, University of Messina, Viale G. Palatucci 13, 98168 Messina, Italy; arenak@unime.it
* Correspondence: cacciolaf@unime.it

1. Introduction

Polyphenolic compounds have long been recognized as one of the most structurally diverse and biologically relevant classes of secondary metabolites in plant-derived foods and natural products [1]. Their importance extends far beyond their classical role as antioxidants, encompassing a broad spectrum of biological, technological, and ecological functions [2]. In food systems, polyphenols contribute to sensory attributes, oxidative stability, and shelf life; in biological systems, they interact with cellular pathways, enzymes, and microbial communities; and in broader sustainability frameworks, they represent key molecular targets for the valorization of agricultural resources and by-products [3]. Against this background, this Special Issue of Molecules entitled "Exploring Bioactive Polyphenolic Compounds in Food and Natural Real-World Samples II" was conceived as a continuation of a previous collection, with the explicit aim of advancing polyphenol research beyond simplified models and toward realistic matrices, scalable processes, and biologically meaningful contexts.

The motivation behind this Special Issue arises from a growing consensus within the scientific community that isolated-compound approaches alone do not full capture the functional properties of polyphenols. While reductionist studies remain indispensable for mechanistic understanding, there is an increasing need to investigate polyphenols as they occur in complex food matrices, plant tissues, and natural samples, where interactions with other constituents profoundly influence their chemical behavior and biological effects [4]. The contributions assembled in this Special Issue respond directly to this challenge by focusing on real-world samples, underutilized materials, and technologically relevant processes to provide a more holistic perspective on polyphenolic compounds and their applications.

2. Overview of Published Articles

Across the literature, a central and recurring theme is the sustainable sourcing and valorization of polyphenols from non-conventional or secondary materials [5]. Several contributions demonstrate that plant parts traditionally regarded as waste, such as oil press-cakes (Contribution 1), fruit husks (Contribution 2), and distillation residues (Contribution 3), can serve as rich reservoirs of phenolic compounds with significant bioactive potential. By combining detailed chemical profiling with functional evaluation, the authors of these studies challenge conventional perceptions of value within food and agro-industrial chains. For instance, the investigation of press-cake powders derived from oilseed processing illustrates how lipid-depleted matrices can retain concentrated phenolic fractions capable of exerting antioxidant and other biological effects. Similarly, the characterization of phenolics recovered from hydro-distillation by-products highlights the possibility of closing material loops within essential oil production systems. Collectively, these works align strongly with the principles of a circular economy, emphasizing resource efficiency, waste minimization, and the creation of added value through molecular-level insight.

Methodological innovation in extraction and processing constitutes another major pillar of this Special Issue. The efficient recovery of polyphenols from complex matrices remains a critical challenge, particularly when scalability, energy consumption, and compound stability must be considered simultaneously (Contribution 4). The application of microwave-assisted extraction, the optimization of drying conditions, and the development of scalable extraction protocols exemplify how process parameters can be fine-tuned to maximize phenolic yield while preserving chemical integrity. Importantly, these studies move beyond purely analytical optimization by addressing technological feasibility and process robustness, thereby narrowing the gap between laboratory research and industrial application. The explicit consideration of factors such as drying temperature and extraction scale underscores the sensitivity of polyphenolic compounds to processing conditions and reinforces the need for carefully designed workflows in both research and production settings.

High-level analytical characterization remains a foundational element of polyphenol research, and the contributions to this Special Issue showcase the power of advanced chromatographic and spectrometric techniques in unraveling complex phenolic profiles. Detailed qualitative and quantitative analyses enable not only the identification of well-established phenolic compounds but also the discovery of previously unreported molecules in wild or less-studied plant species. Comparative studies involving different species, growth conditions, or plant parts further reveal how genetic, environmental, and physiological factors shape polyphenol composition (Contribution 5). These findings underscore the vast and incompletely mapped chemical diversity of polyphenols in nature and highlight the importance of continued methodological refinement to ensure accurate, reproducible, and meaningful characterization.

Beyond compositional analysis, a significant number of contributions integrate chemical data with biological and functional assessments, reflecting an important shift toward analyzing structure–function relationships. Antioxidant activity, antiglycation effects, and metabolic modulation are evaluated using in vitro and ex vivo models, providing insight into the potential health relevance of polyphenol-rich materials. In particular, studies addressing oxidative stress and glycation processes in the context of hyperglycemia emphasize the relevance of polyphenols for metabolic health and chronic disease prevention (Contribution 6). These investigations demonstrate that polyphenols should not be regarded as passive radical scavengers but rather as dynamic molecular actors capable of influencing complex biochemical networks. By linking phenolic composition to functional outcomes, the chosen published works strengthen the scientific basis for the rational development of functional foods and nutraceutical ingredients.

One of the most contemporary and rapidly evolving aspects addressed in this Special Issue is the interaction between polyphenols and gut microbiota (Contribution 7). Growing evidence suggests that many polyphenols exert their biological effects indirectly, through the modulation of microbial composition and activity, as well as through the generation of bioactive metabolites. The investigation of polyphenol-rich extracts in relation to both healthy and diseased gut microbiota provides valuable insight into these complex interactions. Such studies highlight the bidirectional relationship between dietary polyphenols and microbial ecosystems, where polyphenols influence microbial functionality and, in turn, microbial metabolism shapes polyphenol bioavailability and activity. This emerging

https://doi.org/10.3390/molecules31030537

paradigm adds an additional layer of complexity to polyphenol research and encourages future research involving personalized nutrition and microbiota-targeted interventions.

Plant physiology and agronomic practices have also emerged as influential determinants of polyphenol content and composition. Preliminary investigations into elicitation strategies demonstrate that controlled stress conditions can stimulate phenolic accumulation in plant tissues, offering promising avenues for enhancing the functional value of agricultural products at the source (Contribution 8). Although exploratory, these studies suggest that polyphenol optimization need not be limited to post-harvest processing but can be integrated into cultivation strategies, thereby aligning agricultural practices with nutritional and functional objectives. Such approaches may prove particularly valuable in the context of sustainable agriculture and climate-resilient crop production.

From a broader perspective, the contributions to this Special Issue collectively illustrate the multidimensional nature of contemporary polyphenol research. Advances in analytical chemistry, processing technologies, biological evaluation, and system-level understanding are increasingly interconnected, and progress in one area often depends on developments in another. At the same time, the authors of these studies bring attention to persistent challenges that continue to shape the field. These include the need for standardized analytical protocols to facilitate comparability between studies, improved models for assessing bioaccessibility and metabolism, and greater integration of chemical and biological data to support evidence-based health claims.

Interdisciplinary collaboration emerges as a critical driver of progress in this area. Chemists, food scientists, biologists, nutritionists, and technologists must work together to translate molecular insights into practical applications that are safe, effective, and sustainable. Future research will likely benefit from the incorporation of multi-omics approaches, advanced computational tools, and in vivo models capable of capturing the complexity of polyphenol–matrix–host interactions. Such integrative strategies will be essential for moving beyond descriptive studies toward predictive and application-oriented frameworks.

3. Conclusions

This Special Issue titled "Exploring Bioactive Polyphenolic Compounds in Food and Natural Real-World Samples II" provides a comprehensive and timely overview of current research directions in this vibrant field. By emphasizing real-world matrices, sustainable sourcing, methodological rigor, and biological relevance, the collected contributions advance our understanding of polyphenolic compounds and their multifaceted roles in food systems and human health. The guest editors sincerely thank all the authors for their high-quality contributions and commitment to advancing research on polyphenols. We also express our gratitude to the reviewers for their thoughtful and constructive evaluations and to the Molecules editorial team for their professional support throughout the publication process. It is our hope that this Special Issue will serve as a valuable reference point and inspire further interdisciplinary research into the chemistry, functionality, and applications of bioactive polyphenolic compounds.

Author Contributions: All authors contributed to the writing of this manuscript. All authors have read and agreed to the published version of the manuscript.

Funding: This editorial received no external funding.

Conflicts of Interest: The authors declare no conflicts of interest.

https://doi.org/10.3390/molecules31030537

List of Contributions

1. Guarrasi, V.; Prandi, B.; Tedeschi, T.; Chiancone, B.; Di Fazio, A.; Barbieri, R.; Baroni, D.; Soto-Vásquez, M.R.; Vilasi, S.; Falco, F.; Cirlini, M.; Paredes-López, D. Sacha Inchi (*Plukenetia volubilis* L.) Oil Press-Cake Powder: Chemical Characterization and In Vitro Bioactivity for Sustainable Applications. *Molecules* **2026**, *31*, 117.
2. Gomez, P.; Reyes, C.; Figueroa, J.G. Microwave-Assisted Extraction of Phenolic Compounds from Cocoa Pod Husk: Process Optimization and Impact of Drying Temperature on Bioactive Recovery. *Molecules* **2025**, *30*, 3497.
3. Pérez-Magariño, S.; Bueno-Herrera, M.; Asensio-S.-Manzanera, C. Characterization of Bioactive Phenolic Compounds Extracted from Hydro-Distillation By-Products of Spanish Lamiaceae Plants. *Molecules* **2024**, *29*, 5285.
4. Costa, D.; Vasantha Rupasinghe, H.P. Development of a Scalable Extraction Process for Anthocyanins of Haskap Berry (*Lonicera caerulea*). *Molecules* **2025**, *30*, 1071.
5. Bota, V.B.; Oláh, N.-K.; Chise, E.; Burtescu, R.-F.; Pripon Furtună, F.-R.; Ivănescu, L.C.; Zamfirache, M.-M.; Máthé, E.; Turcus, V. A Comparative Analysis of the Polyphenolic Content and Identification of New Compounds from *Oenothera biennis* L. Species from the Wild Flora. *Molecules* **2025**, *30*, 4059.
6. El Mahdi, N.B.; Lemée, L.; Remaury, Q.B.; Eloy, L.; Nhiri, N.; Lakhssassi, N.; Cacciola, F.; Nhiri, M. Potential of *Sorghum* Seeds in Alleviating Hyperglycemia, Oxidative Stress, and Glycation Damage. *Molecules* **2024**, *29*, 3445.
7. Buccato, D.G.; Delgado-Osorio, A.; De Lellis, L.F.; Morone, M.V.; Ullah, H.; Izzo, L.; Lombardi, S.; Di Minno, A.; Riccioni, C.V.; Moriki, D.; Rufián-Henares, J.A.; Daglia, M. Exploring the Influence of a Pomegranate Extract on the Functionality of Healthy and Diseased Human Gut Microbiota:An In Vitro Study. *Molecules* **2025**, *30*, 1634.
8. Salas-Arias, K.; Irías-Mata, A.; Sánchez-Calvo, L.; Brenes-Zárate, M.F.; Abdelnour-Esquivel, A.; Villalta-Romero, F.; Calvo-Castro, L.A. Eliciting Polyphenols in Strawberry Leaves: Preliminary Experiments in *Fragaria* × *ananassa* cv. Festival. *Molecules* **2024**, *29*, 2467.

References

1. Zagoskina, N.V.; Zubova, M.Y.; Nechaeva, T.L.; Kazantseva, V.V.; Goncharuk, E.A.; Katanskaya, V.M.; Baranova, E.N.; Aksenova, M.A. Polyphenols in Plants: Structure, Biosynthesis, Abiotic Stress Regulation, and Practical Applications (Review). *Int. J. Mol. Sci.* **2023**, *28*, 13874. [CrossRef] [PubMed]
2. Galic, L.; Loncaric, Z.; Lisjak, M. A Review of Phenolic Compounds: From Biosynthesis and Ecological Roles to Human Health and Nutrition. *Phyton-Int. J. Exp. Bot.* **2025**, *94*, 3297–3318.
3. Bolat, E.; Sarıta¸, S.; Duman, H.; Eker, F.; Akdasçi, E.; Karav, S.; Witkowska, A.M. Polyphenols: Secondary Metabolites with a Biological Impression. *Nutrients* **2024**, *16*, 2550. [CrossRef] [PubMed]
4. de Araújo, F.F.; de Paulo Farias, D.; Neri-Numa, I.A.; Pastore, G.M. Polyphenols and their applications: An approach in food chemistry and innovation potential. *Food Chem.* **2021**, *338*, 127535. [CrossRef] [PubMed]
5. Pagano, I.; Campone, L.; Celano, R.; Piccinelli, A.L.; Rastrelli, L. Green non-conventional techniques for the extraction of polyphenols from agricultural food by-products: A review. *J. Chromatogr. A* **2021**, *1651*, 462295. [CrossRef] [PubMed]

https://doi.org/10.3390/molecules31030537

molecules

MDPI

Article

Sacha Inchi (*Plukenetia volubilis* L.) Oil Press-Cake Powder: Chemical Characterization and In Vitro Bioactivity for Sustainable Applications

Valeria Guarrasi [1,*], Barbara Prandi [2], Tullia Tedeschi [2], Benedetta Chiancone [2], Andrea Di Fazio [2], Raffaella Barbieri [3], Debora Baroni [3], Marilú Roxana Soto-Vásquez [4], Silvia Vilasi [1], Francesca Falco [5], Martina Cirlini [2] and Daniel Paredes-López [6]

1 Institute of Biophysics, National Research Council (CNR), Via Ugo La Malfa 153, 90146 Palermo, Italy; silvia.vilasi@cnr.it
2 Department of Food and Drug, University of Parma, Viale Parco Area delle Scienze 27/A, 43124 Parma, Italy; barbara.prandi@unipr.it (B.P.); tullia.tedeschi@unipr.it (T.T.); benedetta.chiancone@unipr.it (B.C.); martina.cirlini@unipr.it (M.C.)
3 Institute of Biophysics, National Research Council (CNR), Via De Marini 6, 16149 Genova, Italy; raffaella.barbieri@ibf.cnr.it (R.B.); debora.baroni@ibf.cnr.it (D.B.)
4 Facultad de Farmacia y Bioquímica, Universidad Nacional de Trujillo, Av. Juan Pablo II, Trujillo 13011, Peru; msoto@unitru.edu.pe
5 Institute for Marine Biological Resources and Biotechnology (IRBIM)—CNR, L. Vaccara, 91026 Mazara del Vallo, Italy; francesca.falco@cnr.it
6 Department of Animal Science, Universidad Nacional Agraria de la Selva, Carretera Central Km 1.21, Tingo María 10131, Peru; daniel.paredes@unas.edu.pe
* Correspondence: valeria.guarrasi@ibf.cnr.it

Abstract

Sacha inchi (*P. volubilis* L.), an ancient oilseed crop native to the Amazon, is gaining attention for its high nutritional value particularly due to its ω-3-, -6-, -9-rich oil. However, most research has focused mainly on oil characterization, neglecting the potential of its by-products, such as the Sacha inchi oil-press cake (i.e., the solid residue after oil extraction). This study explores the chemical composition of Sacha inchi oil press-cake powder, focusing on fatty acid and amino acid profiles, antinutrient factors, total phenolic content, antioxidant activity, and the bioactivity of its extracts on cellular models. Fatty acid analysis revealed a high proportion of polyunsaturated fatty acids, especially α-linolenic acid (42.52%), making it a valuable resource for health-promoting applications. The protein content was also significant (41.86%), with a balanced amino acid composition, including essential amino acids such as leucine, valine, and isoleucine, which are vital for muscle protein synthesis and energy metabolism, in food and/or feed applications. Antinutritional factors were detected, including saponins (1050.1 ± 1.1 mg/100 g), alkaloids (2.1 ± 0.5 mg/100 g), and tannins (6.2 ± 0.9 mg/100 g). While these phytotoxins could limit their use in food applications, their potential antimicrobial activity highlights promising pharmacological opportunities. Total phenolic content (TPC) and antioxidant activity (AO) were evaluated using two extract mixtures differing in composition and polarity, with the acetone/water/acetic acid solvent (80/19/1 *v*/*v*/*v*) showing the highest antioxidant properties. The extract obtained showed cytotoxic effects against Panc-1 cancer cells, highlighting its potential in nutraceutical and pharmaceutical applications. This study underscores the unexploited potential of Sacha inchi by-products, such as the oil press-cake, as a sustainable resource of bioactive compounds for functional products, supporting circular bio-economy strategies by plant-based waste and local biodiversity valorization.

Molecules **2026**, *31*, 117 https://doi.org/10.3390/molecules31010117

Keywords: antinutritional factors; antioxidant activity; essential amino acids; fatty acids; Sacha inchi oil press-cake; total phenolic content; pharmacological potential

1. Introduction

Sacha inchi (*Plukenetia volubilis* L.), commonly referred to as the Inca peanut, is an ancient oilseed crop with a rich cultural history and promising industrial potential. Native to the Amazonian regions of Peru, it has been cultivated for over 3000 years, with archeological evidence linking its use to pre-Incan civilizations [1–3]. Traditionally consumed by indigenous communities in both raw and roasted forms, Sacha Inchi has gained increasing attention in recent decades due to its exceptional nutritional profile, particularly its high content of polyunsaturated fatty acids and proteins [4,5].

The Sacha Inchi (SI) seeds are especially valued for their oil, which contains α-linolenic acid (ALA, ω-3) and a favorable ratio of ω-3 to ω-6 fatty acids [5,6]. This composition not only makes it attractive for food applications, but also for use in cosmetics and pharmaceuticals, due to its anti-inflammatory and antioxidant properties [7]. Although Sacha Inchi oil has traditionally been employed for medicinal purposes and skincare by indigenous peoples [7–9], its commercial production has increased significantly in the 21st century, with cultivation spreading across South America and Southeast Asia [10,11]. Despite its well-documented health benefits and growing commercial interest, scientific research has focused primarily on the oil.

SI by-products, such as raw seeds, leaves, and, in particular, the oil press-cake, have been underutilized, often discarded as waste or used primarily for animal feed [12,13]. However, recent interest in the circular economy has highlighted the importance of fully exploiting agricultural by-products. Recent research indicates that the oil press-cake, for instance, is rich in protein, fiber, and bioactive compounds; thus, it holds considerable potential for applications in functional foods, nutraceuticals, and pharmaceutical formulations [13,14].

This study was conducted with the aim of characterizing and valorising one of the most important by-products of SI, the oil press-cake, which in this case was processed into powder. Far from being merely a residue of oil extraction, this by-product represents a promising source for sustainable applications. To this end, we analyzed its fatty acid and amino acid profiles, the presence of antinutritional factors, total phenolic content, and antioxidant capacity, in order to obtain a comprehensive overview of its chemical composition for food and/or feed applications. In addition, we provided novel evidence on the effect of the phenolic extract from *SI* oil press-cake powder (SICP) on cell viability, highlighting its potential for pharmaceutical and anti-tumor applications. To our knowledge, no previous work has combined compositional profiling with an evaluation of its selective bioactivity in human cell lines. Here, we provide the first integrated chemical–biological characterization of SIPC, a necessary step for future nutraceutical and technological developments. Hence, this research offers a comprehensive evaluation of Sacha inchi oil press-cake powder, underscoring its potential as a valuable resource for sustainable product development and as a nutritional and bioactive ingredient within a circular and resource-efficient framework.

2. Results and Discussion

2.1. Fatty Acid Profile

Determining the fatty acid composition allows us to assess the nutritional quality of the residual lipids in the oil press-cake. This information is crucial to evaluate its potential use as a functional ingredient in food or feed formulations, particularly regarding the presence of essential fatty acids and the balance between saturated and unsaturated fats.

https://doi.org/10.3390/molecules31010117

The fatty acid composition of SICP is summarized in Table 1, reporting the individual compounds, their concentrations (μg/g), and relative percentages. The analysis revealed that polyunsaturated fatty acids (PUFAs) represent the predominant class (74.94%), with α-linolenic acid (C18:3, ω-3, ALA) being the most abundant, accounting for over 40% of total fatty acids, followed by linoleic acid (C18:2, ω-6, LA) for over 30% (Table 1)

Table 1. Fatty acids profile (%) of *Sacha inchi* oil press-cake powder.

Fatty Acid		*m/z*	**% ± SD**
Palmitic acid	C16:0	270	5.14 ± 0.37
Palmitoleic acid	C16:1	268	0.09 ± 0.06
Margaric acid	C17:0	284	0.15 ± 0.01
Margaroleic acid	C17:1	282	0.06 ± 0.01
Stearic acid	C18:0	298	4.57 ± 0.12
Oleic acid	C18:1	296	14.21 ± 1.23
Linoleic acid	C18:2	294	32.43 ± 0.89
Linolenic acid	C18:3	292	42.52 ± 0.66
Arachidic acid	C20:0	326	0.16 ± 0.01
Unsaturated arachidic acid	C20:1	324	0.69 ± 0.00
SFA			10.01 ± 0.26
MUFA			15.05 ± 1.30
PUFA			74.94 ± 1.55

Results are expressed as the mean ± standard deviation of three replicates, and are reported as relative percentages (%). Then the sum of saturated fatty acids (SFA), mono-unsaturated (MUFA) and poly-unsaturated (PUFA) fatty acids is presented.

Monounsaturated fatty acids (MUFAs), mainly represented by oleic acid (C18:1), were detected at moderate levels, contributing to the oxidative stability of the matrix. Saturated fatty acids (SFAs), including palmitic acid (C16:0) and stearic acid (C18:0), were present in lower concentrations, which is nutritionally favorable.

Overall, α-linolenic acid, linoleic acid, and oleic acid were the most abundant fatty acids, corresponding to the PUFA and MUFA categories, respectively, for SICP. Other identified fatty acids were moreover present in minor amounts (Table 1). These results are consistent with previous studies on Sacha inchi oil, as reported by Alejandro Ruiz et al. [13] and Maurer et al. [15], who analyzed several authentic and commercial Sacha inchi oil samples and found comparable fatty acid compositions. Indeed, numerous studies have examined the fatty acid profile of Sacha inchi oil due to its recognized nutritional value; however, the oil press-cake fraction remains relatively unexplored, making its chemical characterization—particularly its fatty acid composition—of notable interest. The high content of α-linolenic acid (ALA, 42.52 ± 0.66%) and linoleic acid (LA, 32.43%) in the SICP (Sacha inchi oil press-cake powder) samples confirms the nutritional value of this by-product, as these are essential fatty acids that cannot be synthesized by the human body. This finding highlights Sacha inchi oil press-cake as a valuable plant-based source, especially of ω-3 fatty acids [16], with ALA levels surpassing those typically reported for fish oil—the primary source traditionally used for the extraction of ω-3 fatty acids—which range from 11.9% to 35.3% of total fatty acids [17]. This highlights the potential of Sacha inchi oil press-cake powder as an alternative source of essential ω-3 fatty acids for both nutritional and functional applications, including the development of new food and/or feed products.

https://doi.org/10.3390/molecules31010117

2.2. Amino Acid Profile

The analysis of amino acids provides insight into the protein quality and potential biological value of SICP. This helps determine whether it can serve as a high-quality protein source for human consumption or animal nutrition.

Amino acid profiling indicated the presence of substantial quantities of both essential and non-essential amino acids. SICP contained 41.86% total protein, highlighting its high nutritional value. Among the most abundant non-essential amino acids were glycine (5.68%), aspartic acid (5.66%), glutamic acid (5.16%), and arginine (4.90%).

Essential amino acids were also present in appreciable quantities, including leucine (3.56%), valine (2.48%), and isoleucine (2.05%). Despite the overall richness in protein, methionine content was relatively low (0.81%) compared to other amino acids. As an essential amino acid, methionine's lower concentration suggests the importance of dietary diversification to meet recommended daily intake levels. The total protein percentage was 41.86%, slightly lower than reported in previous studies [18]. All results are summarized in Table 2.

Table 2. Amino acids profile (%) of *Sacha Inchi* oil press-cake powder.

Amino Acid	Relative Percentage (% *w*/*w*) ± SD
Glycine	5.68 ± 0.54
Alanine	1.81 ± 0.07
Serine	3.51 ± 0.18
Proline	2.15 ± 0.06
Valine	2.48 ± 0.08
Threonine	2.50 ± 0.16
Isoleucine	2.05 ± 0.09
Leucine	3.56 ± 0.14
Aspartic Acid *	5.66 ± 0.76
Lysine	1.54 ± 0.16
Glutamic Acid **	5.16 ± 0.69
Histidine	1.40 ± 0.17
Phenylalanine	1.48 ± 0.26
Arginine	4.90 ± 0.43
Tyrosine	2.71 ± 0.50
Methionine	0.81 ± 0.10
Cysteine	1.68 ± 0.15
Total Proteins %	41.86 ± 1.64

* The sum of aspartic acid and asparagine; ** The sum of glutamic acid and glutamine.

Considering amino acid content expressed as percentage (% *w*/*w*), the sample contains appreciable levels of essential amino acids, including leucine, valine, threonine, isoleucine, and phenylalanine, which are critical for multiple physiological functions and must be obtained through the diet. The total protein content is substantial, averaging 41.86%, and the amino acid profile is well-balanced, providing a wide spectrum of amino acids necessary for protein synthesis and diverse metabolic processes. Branched-chain amino acids (BCAAs) such as leucine, valine, and isoleucine are present in particularly high quantities, supporting muscle protein synthesis and energy production, which could be a potential application in functional food products, most commonly required by athletes and

https://doi.org/10.3390/molecules31010117

physically active individuals [19]. Non-essential amino acids, involved in neurotransmitter synthesis and energy metabolism, are also abundant, further enhancing the nutritional value of SI.

2.3. Antinutrient Factors

The antinutritional profile of SICP was evaluated to determine the presence and concentration of secondary metabolites with potential functional effects. The results are summarized in Table 3.

Table 3. Antinutritional metabolites of *Sacha inchi* oil press-cake powder.

Chemical Component	mg/100 g DM
Tannins	6.2 ± 0.9 *
Saponins	1050.1 ± 1.1 **
Alkaloids	2.1 ± 0.5 ***

Results are expressed as means ± SD (standard deviation). DM = dry matter of SICP. * Tannins expressed as mg gallic acid equivalents per 100 g DM (mg GAE/100 g DM). ** Saponins expressed as mg *Quillaja saponaria* equivalents per 100 g DM (mg QSE/100 g DM). *** Alkaloids expressed as mg berberine equivalents per 100 g DM (mg BE/100 g DM).

The analysis revealed that SICP contains measurable amounts of saponins, tannins, and alkaloids, which are classified as antinutritional compounds but are also recognized for their bioactive properties. The saponins were found in a concentration of 1050.1 ± 1.1 mg per 100 g SICP (DM), which indicates a significant occurrence. Tannins were detected at 6.2 ± 0.9 mg per 100 g (expressed as tannic acid), while alkaloids were present at 2.1 ± 0.5 mg per 100 g (expressed as berberine).

These findings suggest that, while SICP contains antinutritional metabolites that could limit direct consumption in large amounts, on the other hand these compounds could contribute to the functional properties of this matrix. Saponins, tannins, and alkaloids are known to exert bioactive effects, including antioxidant, antimicrobial, and potential anticancer activities, which may enhance the value of SICP beyond its nutritional profile.

On the other hand, the presence of antinutritional compounds may limit the direct application of SICP in food formulations [20,21]. But recent studies have demonstrated that hydrothermal treatments, including autoclaving (moist heat at 121 °C for 10–30 min) and hot-air sterilization, can significantly reduce various antinutrients. In particular, marked decreases were observed in alkaloids, nitrates, tannins, saponins, phytic acid, and trypsin inhibitors, with alkaloid reductions reaching up to 90% after 20 min of autoclaving. Beyond the reduction in antinutrients, these treatments also resulted in an increase in protein content (up to 70 g/100 g after 30 min of autoclaving), attributable to the concentration of nutrients and the formation of new protein structures following denaturation and aggregation processes [22]. However, these same compounds, tannins, saponins and alkaloids, have been reported to possess multiple pharmacological activities, including antimicrobial and selective cytotoxic effects [23]. The presence of these bioactive secondary metabolites in the current study is consistent with previous reports [20–22], suggesting that SICP could be valorized as a source of bioactive compounds for pharmaceutical or nutraceutical applications, in addition to its nutritional role.

2.4. Total Phenolic Content and Antioxidant Activity

Regarding TPC and AO, the highest values were obtained using the acetone/water/acetic acid (80/19/1 $v/v/v$) extraction mixture (Table 4). This observation is consistent with Chirinos et al. [24], who reported enhanced extraction yields using the same solvent system. The improved extraction efficiency is likely attributable to the content of condensed tannins in Sacha inchi oil press-cake powder, which are more effectively solubilized in acetone.

https://doi.org/10.3390/molecules31010117

Alternative solvents of varying polarities may also extract phenolic compounds, but often require elevated temperatures (>60 °C) to achieve comparable yields [25]. When acetone is employed as solvent, these temperature requirements can be lower, facilitating efficient extraction [25].

Table 4. Total phenolic content (TPC) and antioxidant activity (AO) measured by DPPH, ABTS and FRAP.

Extraction Method	TPC (mg GAE/g)	DPPH (mg TEAC/g)	ABTS (mg TEAC/g)	FRAP (mg TEAC/g)
Ethanol/water (80/20 *v*/*v*)	0.18 ± 0.01 [b]	0.19 ± 0.01 [b]	0.39 ± 0.16	0.84 ± 0.04
Acetone/water/acetic acid (80/19/1 *v*/*v*/*v*)	0.47 ± 0.02 [a]	0.35 ± 0.01 [a]	1.28 ± 0.33	1.13 ± 0.05

Results are expressed as means ± SD: TPC values are reported as gallic acid equivalents (GAE), while antioxidant capacity is reported as Trolox anti-oxidant equivalents (TEAC). Means within a column with different letters are significantly different ($p < 0.01$).

Since the present study aimed primarily to characterize the phenolic content in terms of TPC and AO rather than to optimize extraction yield, future investigations should focus on elucidating the interplay between solvent type and extraction temperature to maximize phenolic recovery from SICP.

Evaluation of phenolic compounds and antioxidant activity further supports the bioactive potential of Sacha inchi oil press-cake, as reported and highlighted in recent study [26]. In our case, the acetone/water/acetic acid extraction yielded the highest TPC and AO, attributable to condensed tannins and other phenolic compounds with known anti-inflammatory and antioxidant properties. In order to study the antioxidant capacity of the extracts obtained, three different tests were applied: DPPH, ABTS, and FRAP assays. This allows for a better investigation of the actual antioxidant capacity of the extracts, as reported by Thuanthong et al. [27] and Cardinali et al. [28]. The different assays applied for AO determination confirmed the superior radical scavenging capacity as the major potential in reducing ferric ions of the acetone-based extract, highlighting the importance of solvent choice and extraction parameters for optimizing phenolic yield and bioactivity. These results are in line with previous studies on SI leaves, where phenolic concentrations strongly depended on the extraction method [29].

2.5. *Influence on Cellular Viability*

For the cytotoxicity assays, the SICP extract selected was that obtained using the acetone/water/acetic acid (80/19/1 *v*/*v*/*v*) extraction mixture, which had yielded the highest TPC and AO in previous analyses. This extract did not induce any significant cytotoxicity in HEK cells, used as the non-tumoral control. Although the solvent alone (S1) exhibited some inherent cytotoxic effects, treatment with the SICP extract resulted in a significantly greater reduction in cell viability in Panc-1 (pancreatic adenocarcinoma) and HepG2 (hepatocellular carcinoma) cells compared to untreated controls (NT). Conversely, no significant changes in cell viability were observed in PC-3 (prostate adenocarcinoma) cells (Figure 1).

These findings suggest that the compounds contained in the SICP extract are bioactive and exert selective cytotoxic effects on specific human cancer cell lines, notably Panc-1 (pancreatic adenocarcinoma) and HepG2 (hepatocellular carcinoma), while having no significant impact on non-tumoral HEK cells or PC-3 prostate adenocarcinoma cells. This selective cytotoxicity, observed both between tumor and non-tumor cells and among different tumor types, indicates that the bioactive compounds may act through mechanisms

https://doi.org/10.3390/molecules31010117

influenced by the specific molecular and functional characteristics of each cell line, such as variations in redox balance, metabolic activity, or the expression of molecular targets involved in cell survival and proliferation. These observations align with previous reports highlighting the antiproliferative activity of specific chemical classes of compounds in SI, particularly flavonoids and tannins, which have been shown to modulate oxidative stress, interfere with inflammatory signaling pathways, and promote apoptosis in malignant cells [8,9,30,31]. The tumor-type-dependent response observed here underscores the need for further investigation to elucidate the molecular mechanisms underlying this selectivity and supports the rationale for continued exploration of SI-derived compounds as promising candidates in cancer research.

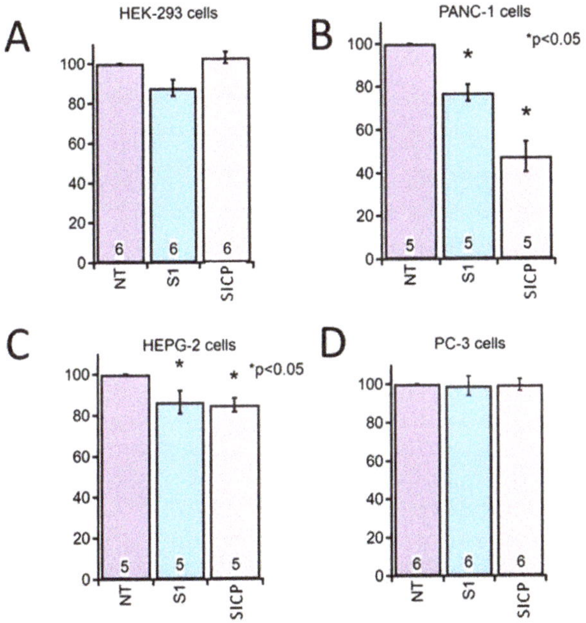

Figure 1. Effect of SICP Extract on Cell Viability in Different Cell Lines. Cell viability was assessed using the MTT assay following 72 h of incubation with SICP extract (SIPC, dilution 1:80) in different human cell lines: (**A**) HEK (human embryonic kidney, non-tumoral control), (**B**) Panc-1 (pancreatic ductal adenocarcinoma), (**C**) HepG2 (hepatocellular carcinoma), and (**D**) PC-3 (prostate adenocarcinoma). NT indicates untreated control cells; S1 indicates cells treated with the solvent mixture alone (acetone/water/acetic acid, 80/19/1 *v*/*v*/*v*), and SIPC indicates cells treated with SICP extracts. Results are expressed as percentage of cell viability relative to untreated controls (NT, 100% viability). Each bar represents the mean ± standard error of the mean (SEM), and the number inside each bar indicates the number of independent biological replicates performed. Statistical significance was evaluated by one-way ANOVA followed by Tukey's post hoc test (* $p < 0.05$ vs. NT control).

3. Materials and Methods

3.1. Raw Material

SI seeds were obtained from agricultural areas in the Province of Picota, San Martín region, Peru (6°54′21.7″ S 76°22′16.1″ W). The seeds were subjected to oil extraction by mechanical pressing at 28 °C, called "cold pressing", a solvent-free process that ensures high-quality oil and produces the solid residue known as SI oil press-cake [32]. This residue was molded into pellets measuring 5 cm in length and 1 cm in diameter (Figure 2A), and subsequently oven-dried at 60 °C for 48 h. The dried pellets were ground in a mill

https://doi.org/10.3390/molecules31010117

(Model 4, Thomas Wiley brand grinder, NJ, USA) to obtain a powder with a particle size of 0.5 mm (Figure 1B). Finally, the powder (SICP) was vacuum-packed and stored at room temperature until analysis.

(A) (B)

Figure 2. (**A**) Pellet-shaped *Sacha inchi* oil press-cake; (**B**) powdered *Sacha inchi* oil press-cake.

3.2. Fatty Acid Profile Determination

To characterize the fatty acid profile of SICP: a 5 g portion of powder was homogenized with n-hexane at room temperature for 30 min to extract the fat fraction. The mixture was filtered using a cellulose paper filter, and the solvent was removed with a rotary evaporator at 40 °C. The extracted fat was weighed, and 100 mg of it was dissolved in 8 mL of hexane. Then, 3 mL of 5% KOH in methanol was added. The mixture was shaken for 1 min to methylate the fatty acids, according to the method described by Annaratone et al. [33].

Subsequently, 0.9 mL of the upper phase of the solution was transferred into a vial, with the addition of 0.1 mL of a tricosanoic acid solution (3000 mg/kg in n-hexane) used as the internal standard. The fatty acids methyl esters (FAME) were analyzed using a Thermo Scientific Trace 1300 gas chromatograph (Thermo Scientific, Waltham, MA, USA) equipped with a Supelcowax (Bellefonte, PA, USA) fused silica capillary column (30 m × 0.25 mm × 0.25 µm film thickness). The system was coupled to a Thermo Scientific Trace ISQ mass spectrometer (Thermo Scientific, Waltham, MA, USA), featuring a single quadrupole analyzer. Helium was used as the carrier gas, and the injection was performed in split mode (split ratio = 1/20) at 280 °C.

The oven temperature program started at 60 °C, increasing to 280 °C at a rate of 20 °C/min after an initial hold of 3 min. The final temperature of 280 °C was maintained for 10 min, with a total run time of 24 min. Fatty acids were identified based on the comparison of their mass spectra with those in the NIST 2020 Mass Spectral Library (National Institute of Standards and Technology, Gaithersburg, MD, USA). Quantification was performed using the tricosanoic acid as the internal reference standard. Results are expressed as the mean of three replicates, calculated as µg/g of dried samples (DM = dry matter), and the relative percentages were reported. Then also the sum of saturated fatty acids (SFA), mono-unsaturated (MUFA) and poly-unsaturated (PUFA) fatty acids is presented.

3.3. Amino Acid Profile Determination

The total amino acid content was determined following the protocol outlined by Prandi et al. [34], with minor modifications. For general amino acid analysis, a 0.5 g sample of SICP was subjected to acid hydrolysis using 6 N hydrochloric acid (HCl) for 23 h at 110 °C (Sigma Aldrich, St. Louis, MO, USA). Nor-leucine (Sigma Aldrich, St. Louis, MO, USA) was added as an internal standard. For sulfur-containing amino acids such as methionine and cysteine, an additional oxidation step using performic acid was carried out prior to acid hydrolysis. Specifically, 50 mg of the sample was treated with freshly prepared

https://doi.org/10.3390/molecules31010117

performic acid (a mixture of 95% formic acid and hydrogen peroxide) and left at 0 °C for 16 h. Performic acid was then removed by adding hydrobromic acid, and samples were dried under nitrogen flow before undergoing acid hydrolysis.

A calibration curve was prepared using a 2.5 mM standard amino acid mixture (Thermo Scientific, Waltham, MA, USA) combined with 1:1 ratio of specific amino acids such as nor-leucine, hydroxyproline, cysteic acid, and methionine sulfone (Sigma Aldrich, St. Louis, MO, USA). This standard solution was diluted to create five concentration points (1.25 mM, 0.625 mM, 0.3125 mM, 0.156 mM, 0.078 mM) for calibration. Samples were derivatized using the AccQ-Fluor reagent kit (Waters, Milford, MA, USA), following the manufacturer's instructions.

Amino acids were analyzed using a UPLC ACQUITY system coupled with an ACQUITY SQ ESI-MS system (Waters, Milford, MA, USA) equipped with a Peptide BEH C18 column (300 Å, 1.7 μm, 2.1 mm 170 × 150 mm) and a VanGuard™ (Agilent, Santa Clara, CA, USA) pre-column (300 Å, 1.7 μm, 2.1 mm × 5 mm). Data acquisition and processing were performed using MassLynx™ V4.0 software (Waters, Milford, MA, USA). Amino acid concentrations (%) were estimated from the calibration curves, using nor-leucine for normalization.

3.4. *Antinutritional Factor Determination*

Antinutritional factors are compounds naturally present in plants, such as tannins, saponins and alkaloids, and may also affect the bioavailability of potentially antioxidant compounds [35]. Their presence can reduce the nutritional value of plant-based foods by binding to essential minerals, inhibiting digestive enzymes, or affecting protein and carbohydrate metabolism [36]. Accurate determination of these compounds is essential for assessing the technological quality of SICP and optimizing subsequent processing methods to ensure its safety and nutritional adequacy in potential food or feed applications.

3.4.1. Tannin Extraction and Quantification

Tannins were quantified according to the Folin–Ciocalteu method by Makkar H.P. [37], with slight modifications. A 0.4 g portion of SICP was mixed with 70 mL of distilled water and placed in an ultrasonic bath for 30 min. The mixture was then filtered using Whatman® No. 2 filter paper, and the filtrate was brought to 100 mL with distilled water. A 1:10 dilution of the extract was prepared by mixing 50 μL of extract with 3950 μL of distilled water and 250 μL of Folin–Ciocalteu reagent. After 2 min, 750 μL of sodium carbonate (20% *v*/*v*) was added. The solution was protected from light and incubated for 1 h at room temperature. Absorbance was measured at 750 nm using a UV–Vis spectrophotometer (UV-1900i, Shimadzu, Tokyo, Japan). Tannin content was determined from a gallic acid calibration curve. Results were expressed as mg gallic acid equivalents per 100 g of dry matter (mg GAE/100 g DM).

3.4.2. Saponin Extraction and Quantification

Saponin quantification was carried out according to the modified method by Guerra et al. [38]. For the preparation of the extract, 1 g of SICP was extracted with 10 mL of ethanol 80% (*v*/*v*) at a solid–liquid ratio of 1:10 (*w*/*v*) under constant agitation for 2 h at room temperature. The solution was then filtered, and the alcoholic extract was used for subsequent dilutions. This extractant was specific to the saponin assay and was not the same solvent mixture used for phenolic extraction. From this alcoholic extract, a 1:10 dilution was prepared using KH_2PO_4 buffer (pH 7.4). A series of eight dilutions (0.3–3.0 mL) was prepared and adjusted to a final volume of 6 mL with distilled water. Subsequently, 6 mL of sodium citrate solution (6%) and 1 mL of erythrocyte suspension were added to each tube. Tubes were incubated at 35 °C for 30 min, centrifuged at 2500 rpm for 5 min, and allowed to stand for 15 min. Absorbance was recorded at 540 nm using

https://doi.org/10.3390/molecules31010117

a UV–Vis spectrophotometer (UV-1900i, Shimadzu, Tokyo, Japan). Quantification was performed using a calibration curve of *Quillaja saponaria* saponins. Results were expressed as mg *Quillaja saponaria* equivalents per 100 g of dry matter (mg QSE/100 g DM).

3.4.3. Alkaloids Extraction and Quantification

A spectrophotometric method based on reaction with bromocresol green (BCG) was used [39]. A 1 g portion of SICP was extracted with 20 mL of absolute methanol three times, using sonication for 20 min at 0 °C. The combined extracts were centrifuged at 4000 rpm for 15 min, filtered, and concentrated in a rotary evaporator. The residue was reconstituted in 20 mL of distilled water, basified to pH 12 with 10% ammonium hydroxide, and transferred to a separation funnel. Alkaloids were extracted into chloroform through repeated liquid–liquid extractions, monitored by thin-layer chromatography using Dragendorff's reagent. The combined chloroform fractions were evaporated to dryness and reacted with BCG reagent. Absorbance was measured at 470 nm, and quantification was performed using a standard curve of berberine chloride (0–10 mg/L); and results were expressed as mg berberine equivalents per 100 g of dry matter (mg BE/100 g DM).

3.5. *Total Phenolic Content and Antioxidant Activity Determination*

For this investigation, two different extraction methods were used and compared with regard to the yield of phenolic compounds and antioxidant activity. The first extraction was performed following the method of Chiancone et al. [40]. An ethanol/water solution (80/20, *v*/*v*) with a solid–liquid ratio of 1:5 was used as the extracting mixture. The extraction was carried out with a shaker (HS 501 digital shaker, IKA-Werke GmbH & Co, Staufen im Breisgau, Germany) at 200 strokes/minute for 2 h at room temperature. For the second extraction, the procedure proposed by Chirinos et al. [24] was followed. A mixture of acetone/water/acetic acid (80/19/1, *v*/*v*/*v*) with a solid–liquid ratio of 1:10 was used to extract the target compounds. The extraction was performed for 60 min at 60 °C with a shaker incubator (SKI4, Shaking Incubator, ARGO LAB, Carpi, Italy). The extracts were centrifuged at 4000 rpm for 10 min at room temperature, and the supernatants were recovered to evaluate the TPC and the AO. Each extraction procedure was repeated twice. All results reported are expressed on dry matter (DM).

3.5.1. Total Phenolic Content Determination

TPC was assessed using Folin–Ciocalteau's phenol reagent, as described by Martelli et al. [41], with slight modifications. 250 µL of the extract was mixed with 1 mL of an aqueous solution of Folin–Ciocalteau's phenol reagent (Sigma-Aldrich, St. Louis, MO, USA) at a ratio of 1:10 (*v*/*v*), followed by the addition of 2 mL of an aqueous solution of sodium carbonate (20%, *w*/*v*). The mixture was then kept in the dark for 30 min. The absorbance at 760 nm was measured using a spectrophotometer (JASCO V-530 spectrophotometer, Jasco, Easton, MD, USA). A calibration curve was generated by utilizing gallic acid as a reference compound across a concentration range of 10–100 mg/kg (5 data points) to determine the polyphenol content in the samples. Each sample extract was analyzed twice, and for enhanced precision, the instrument software was configured to perform three consecutive measurements on each sample. The same procedure was applied for other assays used to determine the antioxidant capacity of the extracts. The results for the total polyphenolic content were expressed as mg of gallic acid equivalents per gram of dry matter (mg GAE/g DM).

3.5.2. Evaluation of Antioxidant Activity by DPPH, ABTS and FRAP Assays

AO of the extracts was assessed using the DPPH (2,2-Diphenyl-1-Picrylhydrazyl) radical scavenging assay, following the method described by Abram et al. [42] with minor adjustments. Specifically, 100 µL of either sample extract or standard solution was com-

https://doi.org/10.3390/molecules31010117

bined with 2.9 mL of a 0.05 mM ethanolic solution of DPPH (Sigma-Aldrich, St. Louis, MO, USA) and left in the dark for 30 min. Subsequently, the absorbance at 517 nm was measured using a JASCO V-530 spectrophotometer (Easton, MD, USA).

To determine the antioxidant capacity, 6-hydroxy-2,5,7,8-tetramethylchroman-2-carboxylic acid (Trolox) (Sigma-Aldrich, St. Louis, MO, USA) was employed as a reference compound, and five different standard solutions (0.1–1 mM) were prepared for constructing the calibration curve. Additionally, a blank containing 100 μL of the extraction solution was analyzed under the same conditions as the samples. The antioxidant capacity was calculated based on the percentage of radical inhibition ($I\%$), using the formula:

$$I\% = 100 * \frac{(Ab_B - Ab_S)}{Ab_B}$$

where Ab_B represents the absorbance of the blank and Ab_S represents the absorbance of the sample/Trolox standard solution. The results were expressed as mg TEAC/mg (Trolox Equivalent Antioxidant Capacity). All analyses were performed in duplicate, with three consecutive measurements conducted on each sample.

Data obtained from the spectrophotometric tests were subjected to one-way Analysis of Variance, considering the factor "Extraction method" with a significance level of 0.05 (IBM SPSS Statistics 26.0 software, SPSS Inc., Chicago, IL, USA).

Similarly, ABTS [2,2-azinobis (3-ethylbenzothiazoline-6-sulfonic acid)] assay was performed following the indications given by Cardinali et al. [28]. The ABTS+ working solution was prepared by mixing ABTS (7 mM) and potassium persulfate (2.45 mM) in distilled water, and diluting this mixture with ethanol in order to obtain an ABTS solution with an absorbance of 0.7 measured at 734 nm. This solution was used to test the extract AO, using 20 μL of sample extract and 1980 μL of ABTS solution. After 30 min of reaction the absorbance values were measured at 734 nm with a JASCO V-530 spectrophotometer (Easton, MD, USA). AO was obtained against Trolox, as already described for DPPH assay.

Finally, Ferric-Ion-Reducing Power (FRAP) was tested, applying the same protocol reported by Cardinali et al. [28]. In particular, a FRAP working solution was created on the basis of a mixture of ferric chloride hexahydrate (20 mM) and 2,4,6-tripyridyl-s-triazine (TPTZ) (10 mM) aqueous solutions, by acidification with acetic acid, and heating at 37 °C. Then, this solution (2850 μL) was allowed to react with sample extracts (150 μL) for 30 min. The absorbance was measured at 593 nm (JASCO V-530 spectrophotometer, Easton, MD, USA) and compared with that of Trolox, using the same calibration curve described for DPPH and ABTS procedures.

3.6. Effect of Phenolic Extract on Cell Viability

To investigate the potential bioactivity of SICP extracts, bioactive compounds were extracted using an acetone/water/acetic acid solvent mixture (80/19/1, *v*/*v*/*v*), which enabled a higher recovery of TPC and enhanced AO in the DPPH assay. After extraction, as previously described, the samples were centrifuged at 4000 rpm for 10 min at room temperature. The resulting supernatant was collected and diluted at a 1:80 ratio for use in cell-based assays.

Panc-1 (human pancreatic ductal adenocarcinoma), HepG2 (human hepatocellular carcinoma), Calu-3 (human lung adenocarcinoma), PC-3 (human prostate adenocarcinoma), and HEK (human embryonic kidney) cell lines were maintained in appropriate growth media following ATCC recommendations. Specifically, Panc-1 and HEK cells were cultured in Dulbecco's Modified Eagle Medium (DMEM), HepG2 and Calu-3 cells in Minimum Essential Medium (MEM) supplemented with 1% non-essential amino acids, and PC-3 cells in DMEM/F-12 (1:1) medium. All media were supplemented with 10% fetal bovine serum

https://doi.org/10.3390/molecules31010117

(FBS), 1% penicillin-streptomycin and 2 mM L-glutamine. Cells were incubated at 37 °C in a humidified atmosphere containing 5% CO_2.

Cell viability was evaluated using the MTT assay, a well-established method to assess cellular metabolic activity [43]. This colorimetric assay measures the enzymatic reduction in the tetrazolium salt MTT (3-[4,5-dimethylthiazol-2-yl]-2,5-diphenyltetrazolium bromide) to insoluble purple formazan crystals by mitochondrial oxidoreductase enzymes, reflecting viable cell function. Cells were incubated with the diluted SICP extract for 72 h. A solvent control (acetone/water/acetic acid, 80/19/1 *v*/*v*/*v*), identical to the extraction solution but without the extract, was included to distinguish specific cytotoxic effects of the phytochemicals. After incubation, MTT reagent was added, and following standard incubation, formazan absorbance was measured spectrophotometrically at 570 nm on a multiplate reader (TriStar2 S LB 942 Multimode Reader, Berthold Technologies, Bad Wildbad, Germany) to determine relative cell viability.

The data are expressed as mean ± standard error of the mean (SEM), calculated from a minimum of five independent biological replicates. Statistical significance of differences among multiple treatment groups was determined using one-way analysis of variance (ANOVA), followed by Tukey's post hoc test for pairwise comparisons. Statistical analyses were performed with IgorPro 9 software (Lake Oswego, OR, USA). Differences were considered statistically significant at a p-value < 0.05.

4. Conclusions

Taken together, these results suggest that SI oil press-cake, often considered a by-product, is a rich source of essential nutrients and bioactive compounds with multifunctional potential. Its ω-3 fatty acids, balanced amino acid profile, phenolic content, antioxidant capacity and selective cytotoxic activity argue for its valorisation as a component of functional food and/or feed, nutraceuticals and pharmaceutical applications, especially when formulated in powder form. Importantly, this work provides the first integrated chemical–biological characterization of SIPC, linking its compositional profile with selective effects on human cell viability. By combining detailed nutrient and antinutrient analysis with multi-line cytotoxicity assays, this study fills a critical gap in the current literature and offers new evidence supporting the biological relevance of SIPC-derived extracts. This integrated perspective strengthens the rationale for the technological and nutraceutical exploitation of this underutilized resource.

Future studies should focus on the isolation and characterization of individual phenolic constituents with antioxidant activity, optimization of extraction techniques for large-scale application and evaluation of bioavailability and safety to fully exploit the industrial potential of Sacha inchi oil press-cake powder.

Author Contributions: Conceptualization, V.G., D.P.-L., M.R.S.-V., B.C. and M.C.; methodology V.G., M.R.S.-V. and M.C.; formal analysis, M.C., B.P., T.T., A.D.F., R.B., D.B. and M.R.S.-V.; investigation, V.G., M.C., B.P., T.T., A.D.F., R.B., D.B. and M.R.S.-V.; resources, D.P.-L.; data curation, V.G., M.C., D.B., S.V. and M.R.S.-V.; writing—original draft preparation, V.G., D.P.-L., D.B., M.R.S.-V. and M.C.; writing—review and editing, V.G., D.P.-L., D.B., B.C., S.V., F.F. and M.C.; supervision, V.G. and M.C.; project administration, V.G. and D.P.-L.; funding acquisition, D.P.-L. All authors have read and agreed to the published version of the manuscript.

Funding: This research was funded by Pro-CIENCIA—Proyectos de Investigación Aplicada 2023-02—PE501082167-2023, project "Torta de *Plukentia Volubilis* (Sacha Inchi) sin metabolitos antinutricionales: una alternativa en la alimentacion de pollos parrilleros" by Universidad Agraria della Selva and by PRIN 2022 PNRR—Funded by the European Union—Next Generation EU, Mission 4 Component 2 Investment 1.1—"Innovative solutions in the wine sector (IN-WINE)", code P2022H573K, CUP B53D23032670001.

https://doi.org/10.3390/molecules31010117

Institutional Review Board Statement: Not applicable.

Informed Consent Statement: Not applicable.

Data Availability Statement: The original contributions presented in this study are included in the article. Further inquiries can be directed to the corresponding author.

Acknowledgments: The authors are grateful to Alessia Provenzano, for her dedicated and precious technical assistance.

Conflicts of Interest: The authors declare no conflicts of interest.

Abbreviations

The following abbreviations are used in this manuscript:

ABTS	2,2′-azinobis(3-ethylbenzothiazoline)-6-sulfonic acid
AO	Antioxidant activity
BCAAs	Branched-chain amino acids
Calu-3	Human lung adenocarcinoma
DM	Dry matter or Sample of SICP
DMEM	Dulbecco's Modified Eagle Medium
DPPH	2,2-diphenyl-1-picrylhydrazyl hydrate
FAME	Fatty acids methyl esters FBS
HepG2	Human hepatocellular carcinoma
HEK	Human embryonic kidney
m/z	Mass-to-charge ratio
MEM	Minimum Essential Medium
MTT	3-(4,5-dimetiltiazol-2-il)-2,5-difeniltetrazolio bromuro assay
MUFA	Mono-unsaturated fatty acids
Panc-1	Human pancreatic ductal adenocarcinoma
PC-3	Human prostate adenocarcinoma
PUFA	Poly-unsaturated fatty acids
SD	Standard deviation
SI	Sacha inchi
SICP	Sacha inchi oil press-cake powder
SFA	Saturated fatty acids
TEAC	Trolox Equivalent Antioxidant Capacity
TPC	Total phenolic content
Trolox	6-hydroxy-2,5,7,8-tetramethylchroman-2-carboxylic acid

References

1. Haros, C.M.; Reguera, M.; Sammán, N.; Paredes-López, O. *Latin-American Seeds: Agronomic, Processing and Health Aspects*; CRC Press: Boca Raton, FL, USA, 2023. [CrossRef]
2. Guillén, M.D.; Ruiz, A.; Cabo, N.; Chirinos, R.; Pascual, G. Characterization of sacha inchi (*Plukenetia volubilis* L.) oil by FTIR spectroscopy and 1H NMR. Comparison with linseed oil. *J. Am. Oil Chem. Soc* **2003**, *80*, 755–762. [CrossRef]
3. Kodahl, N. Sacha inchi (*Plukenetia volubilis* L.)—From lost crop of the Incas to part of the solution to global challenges? *Planta* **2020**, *251*, 80. [CrossRef]
4. Redjeki, S.G.; Hulwana, A.F.; Aulia, R.N.; Maya, I.; Chaerunisaa, A.Y.; Sriwidodo, S. Sacha Inchi (*Plukenetia volubilis*): Potential Bioactivity, Extraction Methods, and Microencapsulation Techniques. *Molecules* **2025**, *30*, 160. [CrossRef] [PubMed]
5. Chirinos, R.; Zuloeta, G.; Pedreschi, R.; Mignolet, E.; Larondelle, Y.; Campos, D. Sacha inchi (*Plukenetia volubilis*): A seed source of polyunsaturated fatty acids, tocopherols, phytosterols, phenolic compounds and antioxidant capacity. *Food Chem.* **2013**, *141*, 1732–1739. [CrossRef] [PubMed]
6. Camargo, R.A.; Medina Vargas, O.J.; Rodríguez Figueroa, F.G.; Moreno Bastidas, L.M. Analysis of physicochemical parameters of Sacha Inchi oil (*Plukenetia Volubilis* L.): A view from production process. *Cienc. En Desarro.* **2025**, *16*, 160–165. [CrossRef]
7. del-Castillo, Á.M.R.; Gonzalez-Aspajo, G.; de Fátima Sánchez-Márquez, M.; Kodahl, N. Ethnobotanical knowledge in the peruvian amazon of the neglected and underutilized crop Sacha Inchi (*Plukenetia volubilis* L.). *Econ. Bot.* **2019**, *73*, 281–287. [CrossRef]

8. Gonzalez-Aspajo, G.; Belkhelfa, H.; Haddioui-Hbabi, L.; Bourdy, G.; Deharo, E. Sacha Inchi Oil (*Plukenetia volubilis* L.), e fect on adherence of Staphylococus aureus to human skin explant and keratinocytes in vitro. *J. Ethnopharmacol.* **2015**, *17* 330–334. [CrossRef]
9. Wang, S.; Zhu, F.; Kakuda, Y. Sacha inchi (*Plukenetia volubilis* L.): Nutritional composition, biological activity, and uses. *Food Chen* **2018**, *265*, 316–328. [CrossRef]
10. Kodahl, L.; Sørensen, J. Sacha inchi (*Plukenetia volubilis* L.) is an underutilized crop with a great potential. *Agronomy* **2021**, *1* 1066. [CrossRef]
11. Maya, I.; Winardi, D.O.; Amalia, E.; Mita, S.R.; Kusumawulan, C.K.; Putriana, N.A.; Sriwidodo, S. Physicochemical characteristics fatty acid profile, and in vitro antioxidant activity evaluation of sacha inchi seed oil from Indonesia. *Cosmetics* **2023**, *1C* 171. [CrossRef]
12. Chen, X.; Wang, K.; Zhang, X. Hypolipidemic activity of protein hydrolysate from Sacha Inchi Meal. *Waste Biomass Valoriz.* **2023** *14*, 2707–2714. [CrossRef]
13. Alejandro Ruiz, F.E.; Ortega Jácome, J.F.; Mora, J.R.; Landázuri, A.C.; Vásconez Duchicela, P.; Vásconez Espinoza, J.; Beltrán Ayala, P.; Andrade-Cuvi, M.J.; Alvarez-Suarez, J.M. Comprehensive characterization and valorization potential of Amazonian Sacha inchi (*Plukenetia volubilis* L.) seeds, oil, and oilcake by-products for sustainable food applications. *Front. Nutr.* **2025**, *12*, 1597300. [CrossRef]
14. Torres-Sánchez, E.; Martínez-Villaluenga, C.; Paterson, S.; Hernández-Ledesma, B.; Gutiérrez, L.F. Antidiabetic and Immunomodulatory Properties of Peptide Fractions from Sacha Inchi Oil Press-Cake. *Foods* **2025**, *14*, 1231. [CrossRef]
15. Maurer, N.E.; Hatta-Sakoda, B.; Pascual-Chagman, G.; Rodriguez-Saona, L.E. Characterization and authentication of a novel vegetable source of omega-3 fatty acids, sacha inchi (*Plukenetia volubilis* L.) oil. *Food Chem.* **2012**, *134*, 1173–1180. [CrossRef]
16. Gutiérrez, L.F.; Rosada, L.M.; Jiménez, A. Chemical composition of Sacha Inchi (*Plukenetia volubilis* L.) seeds and characteristics of their lipid fraction. *Grasas Y Aceites* **2011**, *62*, 76–83. [CrossRef]
17. Rubio-Rodriguez, N.; Beltran, S.; Jaime, I.; Mde Diego, S.; Sanz, M.T.; Carallido, J.R. Production of omega-3 polyunsaturated fatty acid concentrates: A review. *Innov. Food Sci. Emerg. Technol.* **2010**, *11*, 1–12. [CrossRef]
18. Rawdkuen, S.; Murdayanti, D.; Ketnawa, S.; Phongthai, S. Chemical properties and nutritional factors of pressed-cake from tea and sacha inchi seeds. *Food Biosci.* **2016**, *15*, 64–71. [CrossRef]
19. Sanz, J.M.M.; Navarro, A.N.; García, E.S.; López, I.S. An overview on essential amino acids and branched chain amino acids. In *Nutrition* and Enhanced Sports Performance; Elsevier Inc.: Amsterdam, The Netherlands, 2019; pp. 509–519. [CrossRef]
20. Rodzi, N.A.R.M.; Lee, L.K. Sacha Inchi (*Plukenetia volubilis* L.): Recent insight on phytochemistry, pharmacology, organoleptic, safety and toxicity perspectives. *Heliyon* **2022**, *8*, e10572. [CrossRef] [PubMed]
21. Sánchez, E.G.; Hernández-Ledesma, B.; Gutiérrez, L.F. Sacha inchi oil press-cake: Physicochemical characteristics, food-related applications and biological activity. *Food Rev. Int.* **2021**, *39*, 148–159. [CrossRef]
22. Vera, E.L.; Villacres, E.; Ojeda, C.C.; Moyano, V.G.; Tobar, M.Q.; Quelal, M.V.; Yupangui, Y.Q.; Ruales, J. Evaluation of antinutrients, nutritional, and functional properties in sacha inchi (*Plukenetia volubilis* L) cake treated with hydrothermal processes. *Heliyon* **2024**, *10*, e37291. [CrossRef]
23. Berida, T.I.; Kdimy, A.; Adekunle, Y.A.; Roy, S.; Dada-Adegbola, H.; Sarker, S.D. Plant antibacterials: The challenges and opportunities. *Heliyon* **2024**, *10*, e31145. [CrossRef]
24. Chirinos, R.; Necochea, O.; Pedreschi, R.; Campos, D. Sacha inchi (*Plukenetia volubilis* L.) shell: An alternative source of phenolic compounds and antioxidants. *Int. J. Food Sci. Technol.* **2016**, *51*, 986–993. [CrossRef]
25. Das, A.K.; Islam, M.N.; Faruk, M.O.; Ashaduzzaman, M.; Dungani, R. Review on tannins: Extraction processes, applications and possibilities. *S. Afr. J. Bot.* **2020**, *135*, 58–70. [CrossRef]
26. Kittibunchakul, S.; Hudthagosol, C.; Sanporkha, P.; Sapwarobol, S.; Temviriyanukul, P.; Suttisansanee, U. Evaluation of sacha inchi (*Plukenetia volubilis* L.) by-products as valuable and sustainable sources of health benefits. *Horticulturae* **2022**, *8*, 344. [CrossRef]
27. Thuanthong, A.; Klomklao, S.; Panyo, J.; Zhang, Y. In-Vitro Screenings for Biological and Antioxidant Activities of Aqueous Extract from Sacha Inchi (*Plukenetia volubilis* L.) Husks. *Trends Sci.* **2023**, *20*, 6815. [CrossRef]
28. Cardinali, F.; Belleggia, L.; Reale, A.; Cirlini, M.; Boscaino, F.; Di Renzo, T.; Del Vecchio, L.; Cavalca, N.; Milanović, V.; Garofalo, C.; et al. Exploitation of black olive (*Olea europaea* L. cv. Piantone di Mogliano) pomace for the production of high-value bread. *Foods* **2024**, *13*, 460. [CrossRef]
29. Zakaria, Z.; Sabri, S.N.M.; Anwar, N.Z.R.; Shahidan, N. Effect of Different Solvents Extraction on the Total Phenolic Content and Free Radical Scavenging Activity of Sacha Inchi Leaves Oil. *J. Agrobiotechnol.* **2024**, *15*, 11–22.
30. Nascimento, A.K.L.; Melo-Silveira, R.F.; Dantas-Santos, N.; Fernandes, J.M.; Zucolotto, S.M.; Rocha, H.A.O.; Scortecci, K.C. Antioxidant and antiproliferative activities of leaf extracts from *Plukenetia volubilis* Linneo (Euphorbiaceae). *Evid.-Based Complement. Altern. Med.* **2013**, *2013*, 950272. [CrossRef]
31. Wang, M.; Xiang, Y.H.; Liu, M.; Jiang, S.; Guo, J.Y.; Jin, X.Y.; Sun, H.-F.; Zhang, N.; Wang, Z.-G.; Liu, J.-X. The application prospects of sacha inchi (*Plukenetia volubilis* linneo) in rheumatoid arthritis. *Front. Pharmacol.* **2024**, *15*, 1481272. [CrossRef] [PubMed]

https://doi.org/10.3390/molecules31010117

32. Salimi, Y.K.; Tungadi, R.; Liputo, S.A.; Kilo, A.K.; Rauf, A.N.P.; Dali, I.B. Physicochemical Properties, Fatty Acid and Amino Acid Profiles, and Antioxidant Activity of Sacha Inchi (*Plukenetia volubilis* L.) Seed Oil and Press Cake Extracted Using Hydraulic Press and Soxhlet Methods. *Trends Sci.* **2025**, *22*, 9690. [CrossRef]
33. Annaratone, C.; Caligiani, A.; Cirlini, M.; Toffanin, L.; Palla, G. Sterols, sterol oxides and CLA in typical meat products from pigs fed with different diets. *Czech J. Food Sci.* **2009**, *27*, 219–222. [CrossRef]
34. Prandi, B.; Zurlini, C.; Maria, C.I.; Cutroneo, S.; Di Massimo, M.; Bondi, M.; Brutti, A.; Sforza, S.; Tedeschi, T. Targeting the Nutritional Value of Proteins From Legumes By-Products Through Mild Extraction Technologies. *Front. Nutr.* **2021**, *8*, 695793. [CrossRef]
35. Gemede, H.F.; Ratta, N. Antinutritional factors in plant foods: Potential health benefits and adverse effects. *Int. J. Nutr. Food Sci.* **2014**, *3*, 284–289. [CrossRef]
36. Singh, P.; Pandey, V.K.; Sultan, Z.; Singh, R.; Dar, A.H. Classification, benefits, and applications of various anti-nutritional factors present in edible crops. *J. Agric. Food Res.* **2023**, *14*, 100902. [CrossRef]
37. Makkar, H.P. Treatment of plant material, extraction of tannins, and an overview of tannin assays presented in the manual. In *Quantification of Tannins in Tree and Shrub Foliage: A Laboratory Manual*; Springer Science Business Media: Berlin/Heidelberg, Germany, 2003; pp. 43–48.
38. Guerra, J.; Nogueiras, C.; Delgado, R.; Hernández, O. Determinación cuantitativa de saponinas y azucares reductores del Agave brittoniana T. *Rev. Cuba. De Química* **2001**, *13*, 37–42.
39. Shamsa, F.; Monsef, H.; Ghamooshi, R.; Verdianrizi, M. Spectrophotometric determination of total alkaloids in some Iranian medicinal plants. *J. Appl. Hortic.* **2008**, *32*, 17–20. [CrossRef]
40. Chiancone, B.; Guarrasi, V.; Leto, L.; Del Vecchio, L.; Calani, L.; Ganino, T.; Galaverni, M.; Cirlini, M. Vitro-derived hop (*Humulus lupulus* L.) leaves and roots as source of bioactive compounds: Antioxidant activity and polyphenolic profile. *Plant Cell Tissue Organ Cult. (PCTOC)* **2023**, *153*, 295–306. [CrossRef]
41. Martelli, F.; Cirlini, M.; Lazzi, C.; Neviani, E.; Bernini, V. Edible seaweeds and spirulina extracts for food application: In vitro and in situ evaluation of antimicrobial activity towards foodborne pathogenic bacteria. *Foods* **2020**, *9*, 1442. [CrossRef]
42. Abram, V.; Čeh, B.; Vidmar, M.; Hercezi, M.; Lazić, N.; Bucik, V.; Možina, S.S.; Košir, I.J.; Kač, M.; Demšar, L.; et al. A comparison of antioxidant and antimicrobial activity between hop leaves and hop cones. *Ind. Crops Prod.* **2015**, *64*, 124–134. [CrossRef]
43. Van Meerloo, J.; Kaspers, G.J.; Cloos, J. Cell sensitivity assays: The MTT assay. In *Cancer Cell Culture: Methods and Protocols*; Humana press: Totowa, NJ, USA, 2011; pp. 237–245. [CrossRef]

molecules

MDPI

Article

A Comparative Analysis of the Polyphenolic Content and Identification of New Compounds from *Oenothera biennis* L. Species from the Wild Flora

Viviane Beatrice Bota [1,2,3,*,†], Neli-Kinga Oláh [4,5], Elisabeta Chișe [5], Ramona-Flavia Burtescu [4], Flavia-Roxana Pripon Furtună [4], Lăcrămioara-Carmen Ivănescu [1], Maria-Magdalena Zamfirache [1], Endre Máthé [2,6] and Violeta Turcuș [2,3,*,†]

1 Doctoral School of Biology, Faculty of Biology, "Alexandru Ioan Cuza" University of Iași, 700505 Iasi, Romania; ivanescu@uaic.ro (L.-C.I.); magda@uaic.ro (M.-M.Z.)
2 Faculty of Medicine, "Vasile Goldiș" Western University of Arad, 310414 Arad, Romania; endre.mathe@agr.unideb.hu
3 National Institute for Economic Research "Costin C. Kirițescu", Romanian Academy, Centre for Mountain Economy (CE-MONT), 725700 Suceava, Romania
4 SC PlantExtrakt SRL, Rădaia, 407059 Cluj, Romania; neliolah@yahoo.com (N.-K.O.); ramona.burtescu@plantextrakt.ro (R.-F.B.); flavia.pripon@plantextrakt.ro (F.-R.P.F.)
5 Faculty of Pharmacy, "Vasile Goldiș" Western University of Arad, 310414 Arad, Romania; chise.elisabeta@uvvg.ro
6 Institute of Nutrition, Faculty of Agricultural and Food Sciences and Environmental Management, University of Debrecen, H-4032 Debrecen, Hungary
* Correspondence: viviane.beatrice@gmail.com (V.B.B.); turcus.violeta@uvvg.ro (V.T.)
† These authors contributed equally to this work and are both considered the first author.

Abstract

Oenothera biennis L. is a biennial species native to North America and introduced to Europe in the 17th–18th centuries, used in traditional medicine or as a dietary supplement in various products, as well as in cosmetics and the pharmaceutical industry. In recent decades, oil extracted from seeds has been increasingly used for the treatment of various conditions. In this article, we highlight the polyphenolic content of 2 types of extracts from *O. biennis* species, collected from the wild flora of Romania, from two regions with different altitudes and pedoclimatic conditions (plains and mountains), namely: OHM—hydroalcoholic extract, Macea sample; OHVD—hydroalcoholic extract, Vatra Dornei; OAM—aqueous extract, Macea sample; OAVD—aqueous extract, Vatra Dornei sample. The LC/MS analysis of the whole plant extracts revealed 5 to 14 polyphenols, depending on the sample location and type of extract, out of which 7 flavonoids are newly reported for this species. Climatic parameters were mapped using QGIS. Higher qualitative and quantitative values of polyphenols were observed in the hydroalcoholic extract obtained from individuals collected from the plain area.

Keywords: *Oenothera biennis* L.; evening primrose; phytochemistry; polyphenols; flavonoid; phenolic acid; Romania

1. Introduction

Oenothera biennis L. is a herbaceous species belonging to the Onagraceae Juss family [1]. The common name "evening primrose" refers to the flowers that open at sunset [2], while "fever plant" and "king's cure-all" recall the period when the plant was used to treat fever and considered a universal remedy [3].

The species is endemic to North America, with a range that extends from Mexico to Florida, north to central Ontario and Quebec (Canada), west to North Dakota and Oklahoma, and along the Pacific coast [2]. It was subsequently introduced to all continents except Antarctica [2], being the most common and widespread species in the *Oenothera* subsection [4]. Its spread in Europe originates from two distinct groups of *Oenothera* spp. Native to America [4].

In Romania, *O. biennis* is an alien, pioneer species that grows in sandy areas or on skeletal soils, on the edges of roads and railways, on riverbanks, gravel beds, near walls, in ruderal areas, and rarely in cultivated areas, in the *Dauco-Sisymbrion*, *Melilotion*, and *Onopordetalia* plant associations [5,6].

The earliest uses of the species are recorded in the medicinal and dietary practices of Native Americans. Currently, the most widely used part of the plant is the seeds, from which the oil rich in essential fatty acids, especially gamma-linolenic acid (GLA), is extracted and marketed as a dietary supplement or ingredient in various products (cosmetics, pharmaceuticals, etc.) [7–9].

A comprehensive review of the extant literature was conducted, encompassing 50 studies published prior to October 2025, which focus on the phytochemical composition of *O. biennis* L. (Table S1) [10–60]. Research on this species has mainly centered on the composition of seeds and products derived from them (meal, cakes), germinated seeds/sprouts, and seed oil/evening primrose oil (EPO), accounting for 29 studies. The interest in EPO is due to the presence in its composition of omega-6 essential fatty acids, linoleic acid, and gamma-linolenic acid, along with other bioactive compounds such as triterpenes, phenolic acids, tocopherols, and phytosterols [61]. A smaller number of studies (24) highlight the phytochemical composition of the leaves, roots, aerial parts, and the whole plant, while an initial mention of some glycosylated flavonoids does not mention the plant part utilized. The fractions of free and hydrolyzed phenolic acids (acidic and alkaline) in the herb and seeds, respectively, have been highlighted as containing salicylic acid, p-hydroxybenzoic acid, protocatechuic acid, vanillic acid, gentisic acid, p-coumaric acid, caffeic acid, ferulic acid, p-hydroxyphenylacetic acid, syringic acid, gallic acid, pyrocatechuic acid, 2-hydroxy-4-methoxybenzoic acid, and gamma-resorcinol [23,24]. Flavonoids have been confirmed in the leaves [15,29,62] and whole plant [34]. The hydroethanolic extract from sprouts contains miquelianin, luteolin-7-glucuronide, and quercetin-3-O-glucuronide [59,60]. Other flavonoids isolated from seeds were trihydroxyflavanone derivatives, quercetin-xylopyranoside, kaempferol-glucuronide, quercetin, trimethyleneglycol-digallate, monogalloylglucose, digalloylglucose, protocatechuic acid, procyanidin B, procyanidin trimer, (+)-catechin, methyl-gallate, procyanidin B gallate, catechin-gallate [52]. Tannins have been reported in seed meal [54] and leaves [15], and have recently been confirmed in various extracts from the whole plant [34]. Several authors have identified the presence of ellagitannins—oenothein A and B, oligomers of oenothein—in leaves [12,16–18,63], and in the stem and roots [64]. The presence of oenothein B in the aerial parts was highlighted in the aqueous extract [25], respectively, in the crude extract obtained from filtered and lyophilized methanolic extract [26]. Alkaloids have been reported in the leaves [15] and in aqueous, ethanolic, and ethyl acetate extracts of the whole plant [34]. The seeds and the oil extracted from them contain various sterols, 4-methylsterols [35], dimethylsterols [38], phytosterols [35,42,43], 4-desmethylsterol, erythrodiol, and uvaol [51]. Sterols were also found in leaves [14,15], and the roots present oenotheralanosterol A and B of particular medicinal value [31,32].

Other compounds reported in the composition of *O. biennis* are: cetyl alcohol, paraffin, resin, phlobaphenes, tannins, pentosans, mucilage in leaves [5], 2-methyl-7-oxo-tritetracont-1, 5-dien-21-ol [29], dihydroxyprenylxanthone, cetoletyl diglucoside, oenotheaphytilactone, oenotheraphenoxylactone in the root [31,32], tetralin lactones, dodecenyl-benzene-triol, prenyl-anthracene-diol, acyl-diglucoside, and benzoic acid [33], saponins in aqueous, ethanolic, and ethyl acetate extracts from the whole plant [34]. The hydromethanolic extract from the leaves contains: methyl ester 10-octadecenoic acid, 4H-1-bezopyran-4-one-7-hydroxy-2-(4-hydroxyprenyl), 2,6-bis(1,1-dimethyl)-4-[(4-hydroxy-3,5-dimethylphenyl)methyl]-phenol, 4-methyl-1-(1-methylethyl)-3-cyclohexen-1-ol, 3-Buten-2-one, 4-(2,5,6,6-tetramethyl-2-cyclohexan-1-yl), 4,8,12,16-Tetramethylheptadecan-4-olide, and isopropyl stearate [21]. In addition to sugars and amino acids, Wang et al. (2021) [52] highlight the presence of minerals, oxalic acid, and citric acid in seeds.

The scientific literature highlights the biological action of EPO and GLA obtained from seeds in the treatment of diabetes and related conditions. EPO has shown positive effects on the fatty acid profile in patients with type 1 diabetes, influencing several parameters analyzed by Van Doormaal et al. (1988) [65]. The effect of EPO on the circulatory system is often presented in the context of diabetes mellitus pathology, using various in vivo experimental models [66–77]. Several studies show that EPO increases HDL cholesterol levels and reduces serum triglyceride levels [66,68,69,72,78–81]. Additionally, other parts of the plant have also demonstrated biological activity that is relevant in the treatment of conditions such as cancer, as evidenced by studies conducted over the last decade. A water-soluble polysaccharide with antitumor effects was isolated from the whole plant [82]. Oenotheralanosterol B, isolated from the root in combination with oenotheralanosterol A, has shown in vitro antiproliferative activity on liver, breast, prostate, leukemia, and mouse macrophage cell lines [32]. The extract from the aerial parts has antiproliferative and proapoptotic activity on the human melanoma cell line A357 [27]. In vitro, the stem extract stopped the proliferation and migration of vascular smooth muscle cells by regulating proteins involved in the cell cycle [83].

In this paper, we aim to perform a comparative analysis and biochemical characterization of the hydroalcoholic and aqueous extracts from whole *O. biennis* plants. Considering the increasing interest in how environmental factors influence the phytochemical composition of plants, we collected samples from populations growing in two regions with different ecological conditions (plains and mountains), from Romania's wild flora, and analyzed the LC/MS results in relation to altitude and pedoclimatic factors. Based on GIS spatial analysis, the spatial distribution and variability of climatic parameters collected from both site regions over the two-year development period of the species were mapped.

2. Results

Geographical location and biotope conditions are reported to influence the biochemical composition of plants, resulting in different therapeutic properties [84,85]. Therefore, altitude is an important component that introduces variability and influences the main environmental factors [86–88].

For the present study, the *O. biennis* species was collected from two distinct sites in the wild flora of Romania, located in plain and mountain regions, respectively. The analysis of climatic factors from each site area was conducted over two years, considering the biennial nature of the plant.

The first biological samples were collected from Macea, in western Romania. The site is located at an altitude of 99 m, and belongs to the northern forest-steppe zone with *Quercus robur*. From a biogeographical perspective, it corresponds to the Pannonian region. From a climatic perspective, it corresponds to the temperate-continental zone with oceanic influences [89,90]. The other samples for our study were collected from Vatra Dornei in northeastern Romania at an altitude of 926 m. This site belongs to the nemoral zone, the subzone of beech forests (*Fagus sylvatica*) and mixed beech and coniferous forests

(*Abies alba*). Biogeographically, the area corresponds to the Alpine region, and climatically, it falls within the cold continental zone with Scandinavian-Baltic influences [89,90].

In the Macea area, the average annual temperature was 11.7–12.2 °C in the harvest year (2021), and 12.2–12.7 °C in the first year of growth (2020), according to regional weather stations. Similarly, in the Vatra Dornei area, the average annual temperature was between 0.7 and 9.8 °C in 2021 and between 1.9 and 10.5 °C in 2020 (Figure 1).

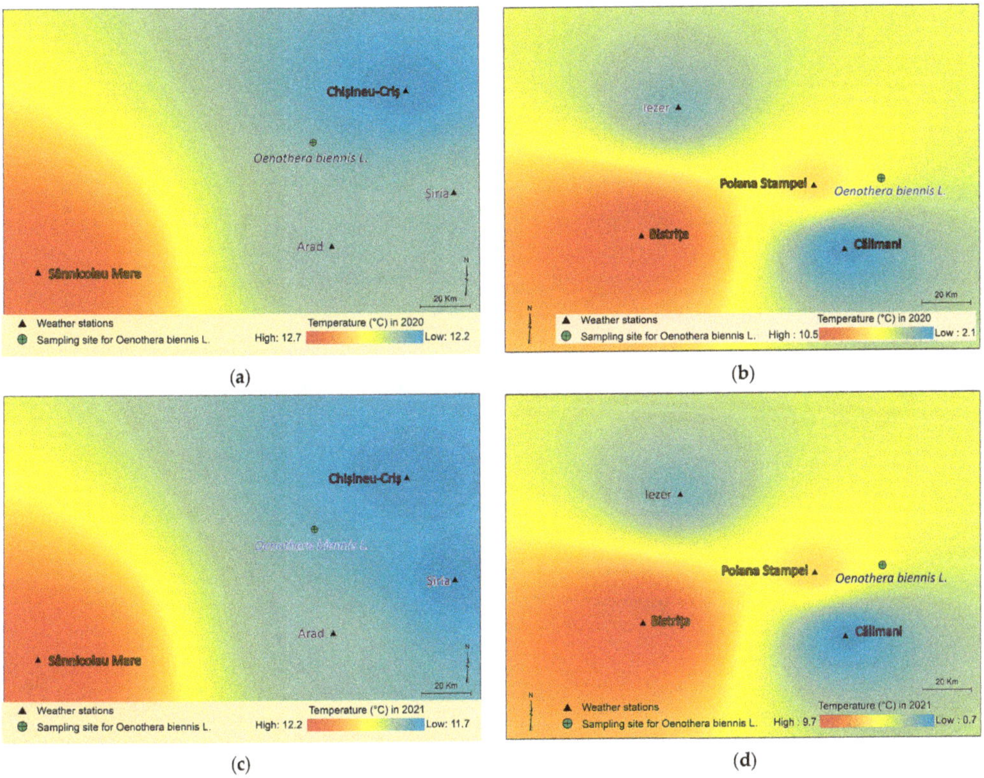

Figure 1. Maps of average annual temperatures based on data from four weather stations near the harvesting locations for the years 2020–2021. (**a**)—Macea 2020, (**b**)—Vatra Dornei 2020, (**c**)—Macea 2021; (**d**)—Vatra Dornei 2021.

In the harvest year of 2021, cumulative precipitation in Macea reached values between 395.3 and 578 mm, with higher values recorded by the stations closest to the locality throughout the year. In 2020, during the early stages of vegetation, cumulative precipitation ranged from 412.9 to 603.8 mm. Higher precipitation values than in the previous year were recorded at the station in Șiria. In the Vatra Dornei area, cumulative precipitation ranged from 757.4 to 1528.1 mm in 2021 and from 691.4 to 1419.5 mm in 2020 (Figure 2).

In terms of soil conditions in the harvesting areas, there are two distinct types of soil: one specific to lowland areas, mainly represented by typical chernozem/black soil with a basic pH, and one specific to mountainous areas, brown soil, which has a characteristic acidic pH. This may also influence the phytochemical profile of the species under study.

Chromatographic analysis of hydroalcoholic and aqueous extracts obtained from whole *O. biennis* plants collected from the wild flora of Romanian plains and mountains revealed the presence of between five and 14 polyphenols, depending on the site of origin and the type of extract. According to the scientific literature to date presented in Table S1, seven flavonoids were identified for the first time in this species.

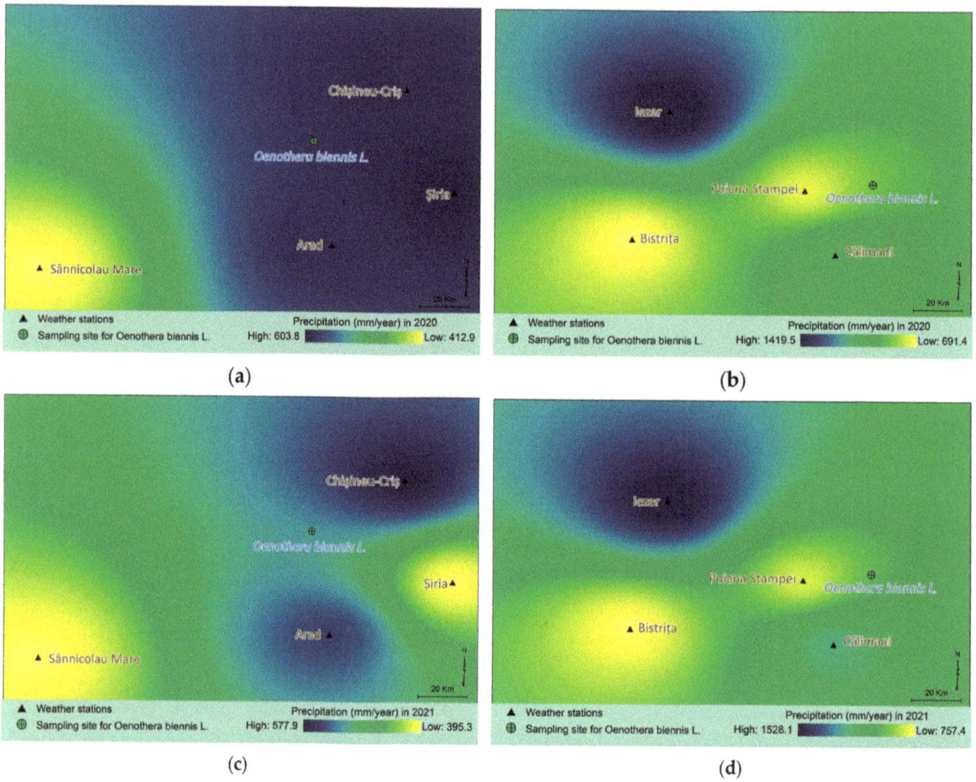

Figure 2. Maps of cumulative annual precipitation based on data from four weather stations near the harvesting locations for the years 2020–2021. (**a**)—Macea 2020, (**b**)—Vatra Dornei 2020, (**c**)—Macea 2021, (**d**)—Vatra Dornei 2021.

The LC/MS analysis of the hydroalcoholic extract allowed the identification and quantification of 14 polyphenols in the sample collected from Vatra Dornei (OHVD), and 13 polyphenols in the sample collected from Macea (OHM) (Table 1).

Table 1. Qualitative and quantitative determination of phenolic acids and flavonoids by LC/MS in hydroalcoholic extracts of *O. biennis* L. from Macea (OHM) and Vatra Dornei (OHVD), in accordance with the standard calibration curves of standards.

Sample Compounds	OHM Sample [μg/g Dry Plant Weight]	OHVD Sample [μg/g Dry Plant Weight]
Caffeic acid	112 ± 2.8	91 ± 2.4
Chlorogenic acid	14.6 ± 0.8	40 ± 1.0
Salicylic acid	347 ± 8.7	199 ± 4.9
Gallic acid	109 ± 2.7	61 ± 1.6
Apigenin	<QL	10 ± 0.6
Chrysin	19 ± 1.1	12 ± 0.7
Esculetin	1632 ± 38.5	1959 ± 46.3
Hesperetin	45.4 ± 1.3	21± 1.2
Hyperoside	732 ± 18.3	116 ± 2.9

Table 1. *Cont.*

Sample Compounds	OHM Sample [μg/g Dry Plant Weight]	OHVD Sample [μg/g Dry Plant Weight]
Kaempferol	16.1 ± 0.9	53 ± 1.4
Luteolin	55.7 ± 1.5	77 ± 2.0
Naringenin	131 ± 3.2	154 ± 3.8
Quercetin	642 ± 16.0	209 ± 5.2
Rutoside	186 ± 4.6	202 ± 5.0

Abbreviations: OHM sample—Hydroalcoholic extract, Macea sample; OHVD sample—Hydroalcoholic extract, Vatra Dornei sample; <QL = compound below quantification limit. Note: Values represent the mean ± standard deviations of three independent measurements of replicate injections of the same extract.

Quantitatively, the main compound in the hydroalcoholic extract was esculetin, in both sites, with the maximum concentration being in the mountain sample (OHVD)—1959 ± 46.3 μg/g dry plant weight. Additionally, OHVD recorded higher concentrations of chlorogenic acid, kaempferol, luteolin, naringenin, and rutoside compared to the plain sample (OHM) (Figure 3). In OHM, other quantitatively important compounds include hyperoside, present at a concentration approximately six times higher than OHVD, quercetin, present at a concentration almost three times higher, and salicylic acid (Figure 3). Apigenin was detected only in the OHVD extract, with a concentration of 10 ± 0.6 μg/g dry plant weight.

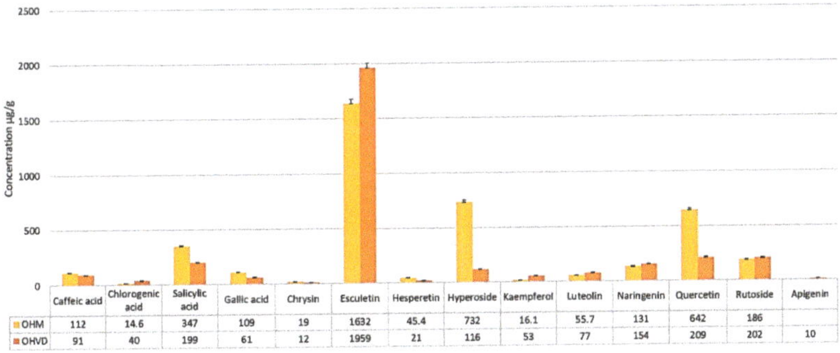

Figure 3. Graphical representation of phenolic acids and flavonoids identified in the hydroalcoholic extract of *O. biennis* collected from Macea (OHM) and Vatra Dornei (OHVD).

A total of 10 polyphenols were identified and quantified in the aqueous extract from the whole *O. biennis* plants in the Macea sample, and 5 in the Vatra Dornei sample. Overall, the quantities were lower than those found in the hydroalcoholic extract, as outlined in Table 2 and Figures 4 and 5.

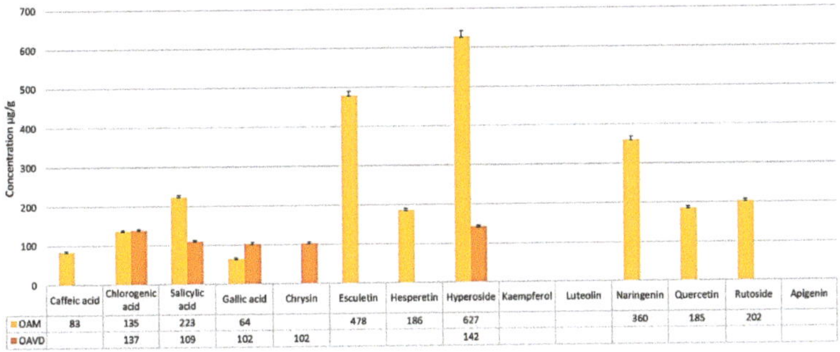

Figure 4. Graphical representation of phenolic acids and flavonoids identified in the aqueous extract of *O. biennis* collected from Macea (OAM) and Vatra Dornei (OAVD).

Table 2. Qualitative and quantitative determination of phenolic acids and flavonoids by LC/MS in aqueous extracts of *O. biennis* L. from Macea (OAM) and Vatra Dornei (OAVD), in accordance with the standard calibration curves of standards.

Sample Compounds	OAM Sample [μg/g Dry Plant Weight]	OAVD Sample [μg/g Dry Plant Weight]
Caffeic acid	83 ± 3.0	<QL
Chlorogenic acid	135 ± 3.3	137 ± 3.2
Salicylic acid	223 ± 5.5	109 ± 2.7
Gallic acid	102 ± 2.6	64 ± 1.7
Apigenin	<QL	<QL
Chrysin	<QL	102 ± 2.6
Esculetin	478 ± 11.9	<QL
Hesperetin	186± 4.6	<QL
Hyperoside	627 ± 15.6	142 ± 3.5
Kaempferol	<QL	<QL
Luteolin	<QL	<QL
Naringenin	360 ± 9.0	<QL
Quercetin	185 ± 4.6	<QL
Rutoside	202 ± 5.0	<QL

Abbreviations: OAM sample—Aqueous extract, Macea sample; OAVD sample—Aqueous extract, Vatra Dornei sample; <QL = compound below quantification limit. Note: Values represent the mean ± standard deviations of three independent measurements of replicate injections of the same extract.

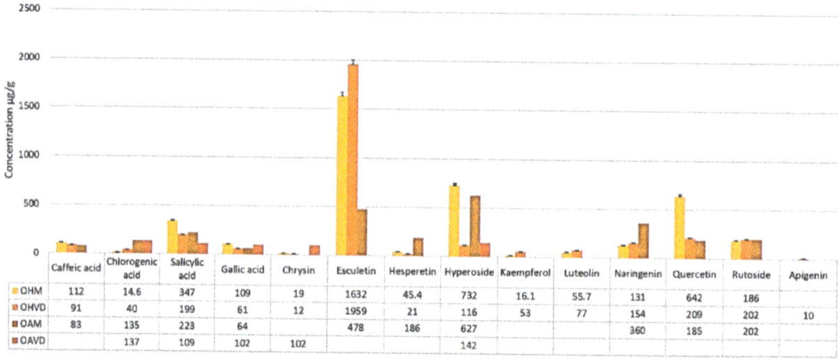

Figure 5. Graphical representation of phenolic acids and flavonoids identified in the hydroalcoholic extract of *O. biennis* collected from Macea (OHM) and Vatra Dornei (OHVD), and in the aqueous extract of the species collected from the same locations (OAM and OAVD, respectively).

The presence of caffeic acid, esculetin, hesperetin, naringenin, quercetin, and rutoside was observed only in the aqueous extract from the plain sample (OAM), while chrysin was observed only in the aqueous extract from the mountain sample (OAVD).

From a quantitative perspective, the major compound in the OAM extract is hyperoside, with a concentration of 627 ± 15.6 μg/g dry plant weight, followed by esculetin, naringenin, salicylic acid, and rutoside. In the OAVD extract, the main compound remains hyperoside, but in a quantity approximately 4 times lower than in the OAM extract, followed by chlorogenic acid (137 ± 3.2 μg/g dry plant weight), salicylic acid (109 ± 2.7 μg/g dry plant weight) and chrysin (102 ± 2.6 μg/g dry plant weight) (Figure 4). However, hyperoside concentration is higher in OAVD than in the OHVD extract (Figure 5).

3. Discussion

Our findings are consistent with previous reports highlighting the influence of environmental and geographical factors on secondary metabolite accumulation in medicinal plants, where altitude, soil composition, and climatic variability have been shown to modulate both the qualitative and quantitative expression of bioactive compounds [91–93].

Regarding climatic factors, temperatures remained relatively constant during the two years of development of the collected plants (2020–2021), with a slight decrease in 2021, in both sites. Concerning precipitation, there was an increase in the mountainous area and a decrease in the plains area in the second year. Therefore, the conditions at the plains collection site were closer to the species' ecological preferences (thermophilic, meso-xerophilic) [90,94].

The LC/MS analysis has revealed up to 14 phenolic acids and flavonoids in the hydroalcoholic extract from *O. biennis* plants collected from the mountain area (OHVD). Only the presence of 2 phenolic acids and 3 flavonoids was retained in the aqueous extract of the same collection site. In regard to the samples from the plain area, 13 of the same polyphenols were confirmed in the hydroalcoholic extract (OHM) and 10 in the aqueous extract (OAM). Seven of the flavonoids identified in this study have not been mentioned in previous publications regarding the phytochemical composition of *O. biennis* species: apigenin, chrysin, esculetin, hesperetin, hyperoside, luteolin, and naringenin. Our present report is primarily aligned with the scientific studies outlined in Table S1 [10–60].

It is important to note that the polyphenols discussed in this paper were identified in extracts obtained from multiple whole plants of the species sampled from two different populations in distinct sites, and that distribution and concentration may also vary with regard to the plant organ. In alignment with previous publications, caffeic acid has been identified in leaves [11,27], seeds [47] and seed meal [55]; chlorogenic acid in seeds [95], neochlorogenic acid in leaves [11]; salicylic acid in herba [23,24] and seeds [24,39]; gallic acid in leaves [11], herba [23,24,27], seeds and seed derivatives [24,39,40,52,54,55,57,60], and sprouts [47]; kaempferol and quercetin in leaves [11,13], herba [26], and seeds [52]; rutoside in herba [27]. Previous reports of compounds identified in this study are also summarized in Table S2.

In addition, there are differences between the hydroalcoholic extract obtained by us from the whole plant in Romania and the sonicated hydroalcoholic extract from the herb obtained by Fecker et al. (2020) [27] from Tunisia. The latter contained higher quantities of caffeic acid, gallic acid, and rutoside, which supports the idea that geographic and pedoclimatic conditions influence this aspect.

While luteolin-7-glucuronide was previously mentioned in the sprout extract, it is a derivative of the luteolin that we are currently reporting in *O. biennis* [96].

The lower quantities of esculetin and flavonoid aglycones such as apigenin, luteolin, and quercetin found in the aqueous extract compared to the alcoholic extract can be explained by their lower solubility in water. In contrast, phenolic acids, chlorogenic acids, and gallic acids have similar solubility in water and ethyl alcohol.

Although the results of the hydroalcoholic extracts indicate a higher number of polyphenols in the mountain sample (OHVD), the overall quantities are lower than in the plain sample (OHM) (Figure 5). According to a 2018 study, the higher number of polyphenols can be explained by the more pronounced stress factors in cold climates [97], *Oenothera biennis* being a species usually associated with a warm climate, similar to that of Macea. However, the higher concentration of polyphenols in the plain sample is in contrast to the observations made by Anstett et al. (2018) [97].

Apigenin was detected in the lowest amount in the OHVD sample (10 μg/g dry plant), and in the OHM sample, it was below the quantification limit, suggesting a possible absence of the compound. This flavonoid is an antioxidant, antimicrobial, and antifungal, playing a role in increasing plant resistance to UV radiation and microbial/fungal infections. In other species, it is present in leaves, fruits, and flowers, and its concentration increases during leaf development [98,99].

The sample from Macea (OHM) contains a higher amount of quercetin (642 ± 16.0 μg/g dry plant) than OHVD (209 ± 5.2 μg/g dry plant), although this flavonoid provides better protection against UV radiation than apigenin [100]. It is possible that in the case of *O. biennis* plants from Vatra Dornei, apigenin production compensates for the lower level of quercetin, and that the change in metabolism and the promotion of the synthesis of other secondary metabolites may be a mechanism of adaptation to mountain stress factors and soil type, associated with the phenomenon of cross-tolerance [84].

Both samples showed a maximum concentration of esculetin, far exceeding that of the other compounds, with the highest amount found in the OHVD sample (1959 μg/g dry plant). This coumarin has numerous pharmacological activities, making it a therapeutic candidate for treating conditions such as diabetes and cancer [101]. According to Tattini et al. (2014) [102], esculetin accumulates in the vacuoles of palisade cells, preferentially in the adaxial portion, playing a role in photoprotection. The synthesis of esculetin coupled with increased peroxidase activity is considered a secondary antioxidant system in conditions of excessive light and depletion of the primary antioxidant system. In the OHM sample, in significant amounts, but less than half the amount of esculetin (1632 ± 38.5 μg/g dry plant), we detected hyperoside (732 ± 18.3 μg/g dry plant), quercetin (642 ± 16.0 μg/g dry plant), and salicylic acid (347 ± 8.7 μg/g dry plant). Hyperoside is a flavonoid-galactoside that plays a role in the plant's response to UV radiation, drought, cold, and salinity stress, as well as reproduction [103]. The role of salicylic acid in plants has been well studied, as it is involved in defense against multiple stress factors and in the reproduction process [104]. In our study, the amount of these three compounds was considerably lower in the OHVD sample, where we note the presence of approximately 200 μg/g dry plant of quercetin, rutoside, and salicylic acid, followed by naringenin (154 μg/g dry plant) and hyperoside (116 μg/g dry plant).

Compared to the hydroalcoholic extract, we observed higher amounts of chlorogenic acid in the aqueous extract (137 ± 3.2—OAVD > 135 ± 3.3—OAM > 40 ± 1.0—OHVD > 14.6 ± 0.81—OHM). The amount of gallic acid was similar between the two types of extract (109 ± 2.7—OHM and 102 ± 2.6—OAM; 64 ± 1.7—OAVD and 61 ± 1.6—OHVD). Chrysin was present in a quantity of 102 μg/g dry plant in OAVD, and below the detection limit in the OAM sample, compared to 12 and 19 μg/g dry plant in the OHVD and OHM extracts, respectively. A similar situation was observed for hesperetin, naringenin, and rutoside, but this time the compounds were below the detection limit in the aqueous extract of the sample from Vatra Dornei (OAVD) and in higher amounts in the OAM sample. Naringenin is a flavonoid capable of inducing tolerance to abiotic stress factors and osmotic stress by regulating nitrogen metabolism and antioxidant effects [105].

As presented above, the newly identified compounds are well-known in the scientific literature for their medicinal properties and role in plant interactions with environmental factors. Therefore, their presence in *O. biennis* L. indicates the therapeutic potential of the species. Moreover, we observed differences in the polyphenolic content of hydroalcoholic and aqueous extracts, as well as between the two sites with characteristic plain and mountain pedoclimatic conditions. The quantitative and qualitative values were higher for hydroalcoholic extracts from *O. biennis* plants originating from low-altitude areas with a warm climate, high average annual temperatures, low cumulative precipitation, and basic soil, such as in the Macea area.

4. Materials and Methods

4.1. Collection of Vegetal Material

The vegetal material consists of whole *Oenothera biennis* L. plants collected from the wild flora of Romania, during their flowering period in August 2021 (in the afternoon), from the following locations:

- Vatra Dornei, Romania; altitude 926 m; GPS coordinates 47° 20′ 30.2136″ N, 25° 19′ 55.0128″ E; the samples were registered with voucher number 1256 in the Herbarium of the Plant Systematics discipline, part of the Department of Biology and Life Sciences at the "Vasile Goldiș" Western University of Arad (Romania) (Figure S1);
- Macea, Romania; altitude 99 m; GPS coordinates 46° 23′ 17.6676″ N, 21° 18′ 7.7934″ E; the samples were registered with voucher number 1257 in the Herbarium of the Plant Systematics discipline, part of the Department of Biology and Life Sciences at the "Vasile Goldiș" Western University of Arad (Romania) (Figure S2).

Several individuals belonging to the same population were collected from each location, with one specimen per population preserved as a voucher, and the rest used for the preparation of extracts. Vouchers correspond to the analyzed lots and were identified by Assoc. Prof. Dr. Turcuș Violeta (Figures S1 and S2). The number of individuals collected per location varied depending on availability and plant biomass. Collection was carried out in compliance with the legislation, research regulations in traditional medicine [106–108], and scientific literature recommendations regarding weather conditions at the time of sampling—warm weather, no precipitation, and ontogenetic stage—at maturity, respectively, during the flowering period [109].

4.2. Preparation of the Hydroalcoholic Extract

For the preparation of the hydroalcoholic extract, the whole plant was dried in the shade at room temperature for 7 days. The dried and crushed plant material was extracted in 70% alcohol at a ratio of 1:5 mass/volume [110,111]. The extract was obtained after 10 days of maceration with daily shaking for 20 min/day at 700–900 rpm, followed by vacuum filtration using W-type analytical filter paper for qualitative determinations, at room temperature, protected from light. The filtrate was brought to the initial volume for all samples. The ethanol used was 70% vol. concentration, obtained from 96% vol. ethanol (Coman Prod, Ilfov, Romania) and purified water (Millipore, Merck, Burlington, MA, USA). The final extraction ratio was 0.2 g dry plant/mL extract. The extraction yield was 20 mg/g dry plant. Concentrations counting the extraction ratio were back-calculated from the extract to the dry plant.

4.3. Preparation of the Aqueous Extract

For the preparation of the aqueous extract, the whole plant was dried for 7 days in the shade at room temperature. The extraction was performed by infusion, at a ratio of 1:10, of dried and crushed plant mixed with hot water (brought to a boil) [112]. The vessel was covered and the mixture was left to infuse for 30 min, then filtered using W-type analytical filter paper for qualitative determinations. The final extraction ratio was 0.1 g dried plant/mL extract. The extraction yield was 10 mg/g dry plant. Concentrations counting the extraction ratio were back-calculated from the extract to the dry plant.

4.4. Liquid Chromatography—Mass Spectrometry Analysis

The LC/MS analysis of plant extracts was performed on a Shimadzu Nexera I LC/MS-8045 UHPLC system (Kyoto, Japan), equipped with a quaternary pump and an autosampler, respectively, with an ESI source and a quadrupole mass spectrometer. Separation was performed on a Luna C18 reverse phase column (150 mm × 4.6 mm × 3 µm, 100 Å) from Phenomenex (Torrance, CA, USA). During analyses, the column was maintained at 40 °C. The mobile phase (Table S3, Figure S3, Supplementary Materials) consisted of a gradient of methanol (Merck, Darmstadt, Germany) and ultrapure water prepared using the Simplicity Ultra Pure Water Purification System (Merck Millipore, Billerica, MA, USA). Formic acid (Merck, Darmstadt, Germany) was used as an organic modifier. Methanol and formic acid were of LC/MS quality. The flow rate used was 0.5 mL/minute, with a total analysis time of 35 min. Detection was performed using a quadrupole mass spectrometer operated with electrospray ionization (ESI), both in negative and positive MRM (multiple reaction monitoring) ionization mode (Table S4). The interface temperature was set at 300 °C. Nitrogen at 35 psi and 10 L/min was used for vaporization and as a carrier gas. The capillary potential was set at +3000 V. The standards are represented by the substances mentioned in Supplementary Materials, Table S5. From each standard, at each concentration, 1 µL was injected. Each sample/extract was diluted 1 mL to 10 mL with methanol, and aliquots of 10 µL were injected into LC/MS. The diluted samples were filtered through 0.45 µm PFTE membrane filters. Identification was performed by comparing the MS spectra and specific transitions between the separated compounds and standards. Identification and quantification were obtained based on the main transition in the MS spectra of the substances. The main transition was used for quantification, and both the main and secondary transitions were used for identification. For quantification purposes, calibration curves were also determined. Table S5 provides the calibration curve equations, their correlation coefficients, and the determined detection and quantification limits. Table S4 includes the retention times and specific transitions of the MS spectra specific to the standards used. Tables S6–S9 show retention times of the standards in parallel with those of the corresponding components separated from the studied extracts. Figures S4–S12 in the Supplementary Materials show the chromatograms, while Figures S13–S19 show the MS spectra.

To determine the matrix effect, standards were added to each extract so that the added concentration was twice that determined in the extract. LC/MS analyses were performed identically to the initial analyses. The same calibration curves were used for the calculations as for the initial determinations. Recovery was determined as the percentage of the ratio of the concentration determined in the sample with standard addition and the standard addition concentration. The recoveries obtained were between 148 and 151%, which corresponds to twice the concentration added to the samples. The matrix effect was determined as the percentage of the ratio between the concentration of the spiked sample and the sum of the concentrations of the unspiked sample and the concentration of the added standard. The percentages obtained range between 99.5 and 100.5%. Therefore, the matrix effect is very low and can be considered insignificant. The determinations for the matrix effect were performed in triplicate, and the recovery and matrix effect percentages mentioned are calculated as the average of the three determinations. The relative standard deviation of the three determinations for each sample and each separate and quantified component is below 10%.

"ND" is defined as "none detected" or "under detection limit". "<QL" is defined as a component that can be identified (at a concentration greater than the detection limit), but which cannot be quantified within a reasonable error limit.

4.5. Spatial Modelling and Thematic Mapping of Climatic Parameters

In order to assess the climatic conditions in the sampling regions (plains and mountains), meteorological data were collected and analyzed from the Meteomanz database [113] from four weather stations close to the collection site for the years 2020–2021, thus encompassing the entire two-year development period of the plant. For the collection site in Macea, the data correspond to those issued by the weather stations in Chișineu-Criș (97 m—46° 31′ N/21° 32′E, Station code 15136); Arad (117 m, 46° 08′ N/21° 21′ E, Station code 15200), Șiria (474 m, 46° 15′ N/21° 39′ E, Station code 15179) and Sânicolaul-Mare (86 m, 46° 04′ N/20° 36′ E, Station code 15199). For collection site in Vatra Dornei, the data correspond to those issued by the weather stations in Poiana Stampei (924 m—47° 19′ N/25° 08′ E, Station code 15069), Călimani-Rețitiș (2021 m, 47° 05′ N/25° 14′ E, Station code 15088), Bistrița (367 m, 47° 08′ N/24° 30′ E, Station code 15085) and Iezer (1787 m, 47° 36′ N/24° 38′ E, Station code 15033).

The spatial distribution of climatic parameters, namely average annual temperature and cumulative precipitation, in the vicinity of the harvesting sites was modelled using QGIS 3.40 "Bratislava" [114]. For this purpose, the Inverse Distance Weighting (IDW) interpolation method [115,116] was applied to datasets obtained from the nearest meteorological stations, a technique that generates estimates for unsampled locations by weighting nearby values inversely with distance, thereby producing continuous raster surfaces that reflect spatial gradients in climate conditions [117,118].

The IDW method was selected as it provides a reliable representation of local climatic variability, particularly in heterogeneous environments such as mountain and lowland areas, where climatic contrasts significantly influence ecological processes. This approach has been shown to support robust environmental modelling and mapping, offering a balance between computational simplicity and interpretative accuracy in climate-ecology interactions [119,120].

The interpolated outputs were subsequently transformed into thematic maps, developed according to established cartographic principles, and integrated into GIS applications that combine environmental, territorial, and socio-cultural data [121–124]. This approach follows key developments in GIScience and applied studies that highlight the usefulness of spatial analysis for understanding landscape dynamics and sustainability issues [125–128].

4.6. Bibliographic Study

The literature review presented in Table S1 was conducted by consulting several types of resources, including original scientific articles, books, book chapters, abstracts, patents, and conference papers, available on major databases such as Web of Science, PubMed, ResearchGate, and Google Scholar. Various combinations of the following keywords were used: *Oenothera biennis*, phytochemistry, chemical composition, polyphenols, apigenin, chrysin, esculetin, hesperetin, hyperoside, luteolin, naringenin, caffeic acid, chlorogenic acid, salicylic acid, gallic acid, kaempferol, quercetin, and rutoside. The bibliographic study encompassed all database search results published before 6 October 2025, including non-English sources (such as Romanian, German, and Japanese with abstract translation). Although review articles were also consulted, only publications reporting original research results were included in Table S1.

4.7. Statistical Analysis

The quantitative values of the compounds identified in the hydroalcoholic and aqueous extracts represent the mean ± standard deviations of three independent measurements of replicate injections of the same extract. Statistical analysis was performed using Excel software from the Microsoft Office package.

5. Conclusions

The analysis of the polyphenolic content of *O. biennis* species collected from two different regions of the wild flora of Romania has revealed quantitative and qualitative differences between the two sites characterized by distinct pedoclimatic factors (plain and mountain), and between the two types of extract (hydroalcoholic and aqueous). We report for the first time in *O. biennis* the presence of 7 flavonoids: apigenin, chrysin, esculetin, hesperetin, hyperoside, luteolin, and naringenin. In general, the polyphenol content of the hydroalcoholic extract is superior to that of the aqueous extract in terms of both quantity and quality. The polyphenol content in the extract obtained from plain area samples (Macea, Romania) is higher compared to that obtained from mountain area samples (Vatra Dornei, Romania). Notable differences were recorded in the content of chlorogenic acid, esculetin, kaempferol, luteolin, and naringenin in the hydroalcoholic extract of *O. biennis* collected from the mountain areas. The amount of gallic acid was similar in both types of extract. Future studies involving multiple sites, larger samples over a longer period, and soil analysis will be necessary to better understand the correlations between ecological factors and polyphenol variability in *O. biennis* plants. Furthermore, a comparative analysis of the polyphenolic content across different plant organs would provide insights into their distribution and primary accumulation sites within this species.

Supplementary Materials: The following supporting information can be downloaded at: https://www.mdpi.com/article/10.3390/molecules30204059/s1, Figure S1: *Oenothera biennis* L. collected from Vatra Dornei. Registered with the voucher number 1256 in the Herbarium of the Plant Systematics discipline, part of the Department of Biology and Life Sciences of the "Vasile Goldiș" Western University of Arad (Romania); Figure S2: *Oenothera biennis* L. collected from Macea. Registered with the voucher number 1257 in the Herbarium of the Plant Systematics discipline, part of the Department of Biology and Life Sciences of the "Vasile Goldiș" Western University of Arad (Romania); Figure S3: Graphical representation of the mobile phase gradient of LC/MS analysis; Figures S4–S6: LC/MS chromatogram of the hydroalcoholic extract obtained from *O. biennis* L. plants from Vatra Dornei (OHVD); Figures S7–S9: LC/MS chromatogram of the hydroalcoholic extract obtained from *O. biennis* L. plants from Macea (OHM); Figure S10: LC/MS chromatogram of the aqueous extract obtained from *O. biennis* L. plants from Vatra Dornei (OAVD); Figures S11 and S12. LC/MS chromatogram of the aqueous extract obtained from *O. biennis* L. plants from Macea (OAM); Figures S13 and S14: MS spectra of separated compounds from the hydroalcoholic extract of *O. biennis* L. plants from Macea (OHM); Figures S15 and S16: MS spectra of separated compounds from the hydroalcoholic extract of *O. biennis* L. plants from Vatra Dornei (OHVD); Figures S17 and S18: MS spectra of separated compounds from the aqueous extract of *O. biennis* L. plants from Macea (OAM); Figure S19: MS spectra of separated compounds from the aqueous extract of *O. biennis* L. plants from Vatra Dornei (OAVD). Table S1: Phytochemical composition of *O. biennis* L.—summarized results of original articles sorted by chronological appearance, plant part, and extraction method; Table S2: Prior reports in *Oenothera biennis* L. species, by plant part and extract type, of compounds identified in the present study; Table S3: Mobile phase composition for LC/MS analysis; Table S4: The standards for chromatographic and spectral data; Table S5: The standards used for LC/MS analyses; Table S6: Retention times and main transitions for each compound identified in the hydroalcoholic extract of *O. biennis* from Macea (OHM); Table S7: Retention times and main transitions for each compound identified in the hydroalcoholic extract *of O. biennis* from Vatra Dornei (OHVD); Table S8: Retention times and main transitions for each compound identified in the aqueous extract of *O. biennis* from Macea (OAM); Table S9: Retention times and main transitions for each compound identified in the aqueous extract of *O. biennis* from Vatra Dornei (OAVD).

Author Contributions: Conceptualization, M.-M.Z., E.M. and V.T.; methodology, V.B.B., N.-K.O., E.C., R.-F.B. and F.-R.P.F.; software, V.B.B., E.C., R.-F.B. and F.-R.P.F.; validation, V.B.B., E.C., R.-F.B. and F.-R.P.F.; formal analysis, V.B.B., N.-K.O., E.C., R.-F.B., F.-R.P.F., L.-C.I., M.-M.Z., E.M. and V.T.; investigation, M.-M.Z. and V.T.; resources, N.-K.O., L.-C.I., M.-M.Z. and V.T.; data curation, V.B.B., N.-K.O. and V.T.; writing—original draft preparation, V.B.B., N.-K.O., E.C., R.-F.B. and F.-R.P.F.; writing—review and editing, N.-K.O., L.-C.I., M.-M.Z., E.M. and V.T.; visualization, V.B.B., N.-K.O., and V.T.; supervision, N.-K.O., M.-M.Z., E.M. and V.T.; project administration, N.-K.O., M.-M.Z.

and V.T.; funding acquisition, N.-K.O., M.-M.Z. and V.T. All authors have read and agreed to the published version of the manuscript.

Funding: This research received no external funding.

Institutional Review Board Statement: Not applicable.

Informed Consent Statement: Not applicable.

Data Availability Statement: The data presented in this study are available within the article and/or Supplementary Materials.

Conflicts of Interest: Authors Neli-Kinga Oláh, Ramona-Flavia Burtescu and Flavia-Roxana Pripon Furtună were employed by the company SC PlantExtrakt SRL. The remaining authors declare that the research was conducted in the absence of any commercial or financial relationships that could be construed as a potential conflict of interest.

References

1. WFO [World Flora Online]: *Oenothera biennis* L. Available online: https://www.worldfloraonline.org/taxon/wfo-0000389564 (accessed on 2 October 2023).
2. Steckel, L.E.; Sosnoskie, L.M.; Steckel, S.J. Common evening-primrose (*Oenothera biennis* L.). *Weed Technol.* **2019**, *33*, 757–760. [CrossRef]
3. Calapai, G.; Minciullo, P.L.; Miroddi, M.; Chinou, I.; Gangemi, S.; Schmidt, R.J. Contact dermatitis as an adverse reaction to some topically used European herbal medicinal products-Part 3: Mentha × piperita-Solanum dulcamara. *Contact Dermat.* **2016**, *74*, 131–144. [CrossRef]
4. Dietrich, W.; Wagner, W.L.; Raven, P.H. *Systematic Botany Monographs*; The American Society of Plant Taxonomists: Laramie, WY, USA, 1997; Volume 50, pp. 20–39.
5. Săvulescu, T. Genul *Oenothera* L. *Oenothera biennis* L. In *Flora Republicae Popularis Romanicae*; Ghișa, E., Grințescu, I., Gușuleac, M., Morariu, I., Nyárády, E.I., Prodan, I., Eds.; Editio Academiae Republicae Popularis Romanicae: Bucharest, Romania, 1957; Volume V, pp. 514–516.
6. Sârbu, I.; Ștefan, N.; Oprea, A.; Bortaș, V. *Plante Vasculare din România: Determinator Ilustrat de Teren*; Editura Victor B Victor: Bucharest, Romania, 2013.
7. Balch, A. Evaluation of Native Evening Primrose for Gamma-Linolenic Acid. Ph.D Thesis, Texas Tech University, Lubbock, TX, USA, August 2000.
8. Kuhn, M.A.; Wiston, D. *Herbal Therapy & Supplements: A Scientific & Traditional Approach*, 2nd ed.; Wolters Kluwer Health | Lippincott Williams & Wilkins: Philadelphia, PA, USA, 2007; pp. 181–185.
9. Stonemetz, D. A Review of the Clinical Efficacy of Evening Primrose. *Holist. Nurs. Pract.* **2008**, *22*, 171–174. [CrossRef]
10. Howard, G.Z.; Mabry, T.J.; Raven, P.H. Distribution of flavonoids in twenty-one species of Oenothera. *Phytochemistry* **1972**, *11*, 289–291. [CrossRef]
11. Zinsmeister, H.D.; Bartl, S. The phenolic compounds of Oenothera. *Phytochemistry* **1971**, *10*, 3129–3132. [CrossRef]
12. Yoshida, T.; Tong, C.; Muneto, M.; Taeko, Y.; Kuzufumi, Y.; Tsutomu, H.; Aya, N.; Takuo, O. Woodfordin D and Oenothein A, Trimeric Hydrolyzable Tannins of Macro-Ring Structure with Antitumor Activity. *Chem. Pharm. Bull.* **1991**, *39*, 1157–1162. [CrossRef] [PubMed]
13. Johnson, M.T.J.; Agrawal, A.A.; Maron, J.L.; Salminen, J.-P. Heritability, covariation and natural selection on 24 traits of common evening primrose (*Oenothera biennis*) from a field experiment. *J. Evol. Biol.* **2009**, *22*, 1295–1307. [CrossRef]
14. Weglarz, Z.; Kosakowska, L.; Pelc, M.; Geszprych, A.; Przybyl, J.L.; Baczek, K. Intraspecific variability of fireweed (*Chamaenerion angustifolium*/L./Scop.) and evening primrose (*Oenothera biennis* L.) in respect of sterol content. *Herba Pol.* **2011**, *57*, 7–15.
15. Lakshman, J.A.; Shasidhar, M.; Ravi Kumar, A. Preliminary phytochemical and toxicity studies of *Oenothera biennis*. *Indian J. Res. Pharm. Biotechnol.* **2014**, *2*, 1114–1118.
16. Anstett, D.N.; Cheval, I.; D'Souza, C.; Salminen, J.-P.; Johnson, M.T.J. Ellagitannins from the Onagraceae Decrease the Performance of Generalist and Specialist Herbivores. *J. Chem. Ecol.* **2019**, *45*, 86–94. [CrossRef]
17. Agrawal, A.A.; Hastings, A.P.; Johnson, M.T.J.; Maron, J.L.; Salminen, J.-P. Insect Herbivores Drive Real-Time Ecological and Evolutionary Change in Plant Populations. *Science* **2012**, *338*, 113–116. [CrossRef] [PubMed]
18. Anstett, D.N.; Ahern, J.R.; Glinos, J.; Nawar, N.; Salminen, J.-P.; Johnson, M.T.J. Can Genetically Based Clines in Plant Defence Explain Greater Herbivory at Higher Latitudes? Edited by Ted Turlings. *Ecol. Lett.* **2015**, *18*, 1376–1386. [CrossRef] [PubMed]
19. Noge, K.; Tamogami, S. Isovaleronitrile co-induced with its precursor, l-leucine, by herbivory in the common evening primrose stimulates foraging behavior of the predatory blue shield bug. *Biosci. Biotechnol. Biochem.* **2018**, *82*, 395–406. [CrossRef]

20. Dong, Z.Y.; Yin, X.S.; Liu, S.L.; Zhang, T.H.; Ren, H.; Tan, H.Y. Ultrasonic-assisted Extraction, Antioxidant Activity and Structura Characterization of Polysaccharides from *Oenothera biennis* L. Leaves. *E3S Web Conf.* **2020**, *189*, 02025. [CrossRef]
21. Patan, A.; Saranya, M.; Vignesh, S.; Bharathi, A.; Vikram, G.; Yuvaraj, P.; Vijeyaanandhi, M. Qualitative Phytochemical Analysi in Determination of Antioxidant Activity of Methanolic Extract of *Oenothera biennis* by GC-MS–A Preliminary Research Stud *Res. J. Pharm. Technol.* **2021**, *14*, 3744–3750. [CrossRef]
22. Mukherjee, K.D.; Kiewitt, I. Formation of gamma-linolenic acid in the higher plant evening primrose (*Oenothera biennis* L. *J. Agric. Food Chem.* **1987**, *35*, 1009–1012. [CrossRef]
23. Krzaczek, T.; Bogucka-Kocka, A. P21 chromatographical analysis of phenolic acids in some species genus *Oenothera* L. *Eu. J. Pharm. Sci.* **1994**, *2*, 117–194.
24. Krzaczek, T.; Bogucka-Kocka, A.; Śnieżko, R. The phenolic acids of some species of the Oenothera L. genus. *Acta Soc. Bot. Po* **1995**, *64*, 41–44. [CrossRef]
25. Granica, S.; Bazylko, A.; Kiss, A.K. Determination of Macrocyclic Ellagitannin Oenothein B in Plant Materials by HPLC-DAD-MS Method Development and Validation. *Phytochem. Anal.* **2012**, *23*, 582–587. [CrossRef]
26. Granica, S.; Czerwinska, M.E.; Piwowarski, J.P.; Ziaja, M.; Kiss, A.K. Chemical composition, antioxidative and anti-inflammator activity of extracts prepared from aerial parts of *Oenothera biennis* L. and *Oenothera paradoxa* Hudziok obtained after seed cultivation. *J. Agric. Food Chem.* **2013**, *61*, 801–810. [CrossRef]
27. Fecker, R.; Buda, V.; Alexa, E.; Avram, S.; Pavel, I.Z.; Muntean, D.; Cocan, I.; Watz, C.; Minda, D.; Dehelean, C.A.; et al Phytochemical and Biological Screening of *Oenothera biennis* L. Hydroalcoholic Extract. *Biomolecules* **2020**, *10*, 818. [CrossRef] [PubMed]
28. Miyazawa, M.; Ikeda, H.; Hiromu, K. The constituents of the essential oil from *Oenothera biennis* L. *Nippon. Nōgeikagaku Kaishi* **1978**, *52*, 449–455. [CrossRef]
29. Shukla, Y.N.; Srivastava, A.; Kumar, S. Aryl, lipid and triterpenoid constituents from *Oenothera biennis*. *Indian J. Chem.* **1999**, *38*, 705–708.
30. Shukla, Y.N.; Ranganathan, T.; Kumar, S.; Srivastava, A.; Khanuja, S.P.S.; Gupta, V.K.; Kumar, S. Antibacterial Composition Comprising Oenostacin from Oenothera biennis. US Patent 6.451.356 B1, 17 September 2002.
31. Ahmad, A.; Singh, D.K.; Fatima, K.; Tandon, S.; Luqman, S. New constituents from the roots of *Oenothera biennis* and their free radical scavenging and ferric reducing activity. *Ind. Crop Prod.* **2014**, *58*, 125–132. [CrossRef]
32. Singh, S.; Dubey, V.; Singh, D.K.; Fatima, K.; Ahmad, A.; Luqman, S. Antiproliferative and antimicrobial efficacy of the compounds isolated from the roots of *Oenothera biennis* L. *J. Pharm. Pharmacol.* **2017**, *69*, 1230–1243. [CrossRef]
33. Sultana, N.S.; Ali, N.M.; Rasool, S. Chemical Constituents From the Roots of *Oenothera biennis* L. *Pharm. Biosci. J.* **2018**, *6*, 29–35. [CrossRef]
34. Ambrin, D.G.; Bakht, J.; Adil, M. Phytotoxic, insecticidal and cytotoxic activities of *Ziziphus mauritiana* var. *spontanea* Edgew. and *Oenothera biennis* L. *Pak. J. Bot.* **2020**, *52*, 2191–2195. [CrossRef]
35. Hudson, B.J.F. Evening primrose (*Oenothera* spp.) oil and seed. *J. Am. Oil Chem. Soc.* **1984**, *61*, 540–543. [CrossRef]
36. Ratnayake, W.M.N.; Matthews, D.G.; Ackman, R.G. Triacylglycerols of evening primrose *Oenothera biennis* seed oil. *J. Am. Oil Chem. Soc.* **1989**, *66*, 966–969. [CrossRef]
37. Court, W.A.; Hendel, J.G.; Pocs, R. Determination of the fatty acids and oil content of evening primrose (*Oenothera biennis* L.). *Food Res. Int.* **1993**, *26*, 181–186. [CrossRef]
38. Ntsourankoua, H.; Artaud, J. Determination and identification of triterpenic alcohols in borage, blackcurrant and evening primrose oils. *Ol. Corps Gras Lipides* **1997**, *4*, 147–151.
39. Zadernowski, R.; Naczk, M.; Nowak-Polakowska, H. Phenolic acids of borage (*Borago officinalis* L.) and evening primrose (*Oenothera biennis* L.). *J. Am. Oil Chem. Soc.* **2002**, *79*, 335–338. [CrossRef]
40. Aitani, M.; Kimura, H.; Yasuhiro, A.; Soyama, H.; Murakami, H.; Zhang, H.L.; Sugishita, T.; Konishi, Y. Effect of an Extract from Evening-Primrose Seeds on Postprandial Blood Glucose Level and Its Active Components. *Nippon Shokuhin Kagaku Kogaku Kaishi* **2003**, *50*, 180–187. [CrossRef]
41. Zaugg, J.; Olivier, P.; Plescher, A.; Bernd, H.; Hamburger, M. Quantitative Analysis of Anti-inflammatory and Radical Scavenging Triterpenoid Esters in Evening Primrose Seeds. *J. Agric. Food Chem.* **2006**, *54*, 6623–6628. [CrossRef] [PubMed]
42. Eskin, N.A.M. Borage and evening primrose oil. *Eur. J. Lipid Sci. Technol.* **2008**, *110*, 651–654. [CrossRef]
43. Nogala-Kalucka, M.; Rudzinska, M.; Zadernowski, R.; Siger, A.; Krzyzostaniak, I. Phytochemical content and antioxidant properties of seeds of unconventional oil plants. *J. Am. Oil Chem. Soc.* **2010**, *87*, 1481–1487. [CrossRef]
44. Montserrat-de la Paz, S.; Fernández-Arche, Á.; Ángel-Martín, M.; García-Giménez, M.D. The sterols isolated from Evening Primrose oil modulate the release of proinflammatory mediators. *Phytomed* **2012**, *19*, 1072–1076. [CrossRef]
45. Montserrat-de la Paz, S.; Fernández-Arche, M.A.; Ángel-Martín, M.; García-Giménez, M.D. Phytochemical characterization of potential nutraceutical ingredients from Evening Primrose oil (*Oenothera biennis* L.). *Phytochem. Lett.* **2013**, *8*, 158–162. [CrossRef]

46. Montserrat-de la Paz, S.; García-Giménez, M.D.; Ángel-Martín, M.; Pérez-Camino, M.C.; Fernández Arche, A. Long-chain fatty alcohols from evening primrose oil inhibit the inflammatory response in murine peritoneal macrophages. *J. Ethnopharmacol.* **2014**, *151*, 131–136. [CrossRef]
47. Pająk, P.; Socha, R.; Broniek, J.; Królikowska, K.; Fortuna, T. Antioxidant properties, phenolic and mineral composition of germinated chia, golden flax, evening primrose, phacelia and fenugreek. *Food Chem.* **2019**, *275*, 69–76. [CrossRef]
48. Zhao, B.; Gong, H.; Li, H.; Zhang, Y.; Deng, J.; Chen, Z. Fatty Acid, Triacylglycerol and Unsaponifiable Matters Profiles and Physicochemical Properties of Chinese Evening Primrose Oil. *J. Oleo Sci.* **2019**, *68*, 719–728. [CrossRef]
49. Ghasemnezhad, A.; Honermeier, B. Seed yield, oil content and fatty acid composition of *Oenothera biennis* L. affected by harvest date and harvest method. *Ind. Crops Prod.* **2007**, *25*, 274–281. [CrossRef]
50. Ghatas, Y.; Mohamed, Y. Influence of some phosphorus sources and biofertilizers (em and phosphorein) on vegetative growth, fixed oil productivity and chemical constituents of *Oenothera biennis* L. plant. *Sci. J. Flowers Ornam. Plants* **2020**, *7*, 247–268. [CrossRef]
51. Zielińska, A.; Kubasiewicz, K.; Wójcicki, K.; Silva, A.M.; Nunes, F.M.; Szalata, M.; Słomski, R.; Eder, P.; Souto, E.B. Two- and Three-Dimensional Spectrofluorimetric Qualitative Analysis of Selected Vegetable Oils for Biomedical Applications. *Molecules* **2020**, *25*, 5608. [CrossRef]
52. Wang, Z.; Wu, Z.; Zuo, G.; Lim, S.S.; Yan, H. Defatted Seeds of *Oenothera biennis* as a Potential Functional Food Ingredient for Diabetes. *Foods* **2021**, *10*, 538. [CrossRef]
53. Sumara, A.; Stachniuk, A.; Trzpil, A.; Bartoszek, A.; Montowska, M.; Fornal, E. LC–MS Metabolomic Profiling of Five Types of Unrefined, Cold-Pressed Seed Oils to Identify Markers to Determine Oil Authenticity and to Test for Oil Adulteration. *Molecules* **2023**, *28*, 4754. [CrossRef]
54. Šardzíková, I.; Schmidt, Š.; Farkaš, P.; Sekretar, S.; Kovac, M. Isolation and Antioxidant Activity of Phenolic Compounds of Evening Primrose (*Oenothera biennis* L.). *Bulletin Potravinárskeho Výskumu (Bull. Food Res.) Roè.* **2002**, *41*, 241–253.
55. Zahradníková, L.; Schmidt, Š.; Sékelyová, Z.; Sekretár, S. Fractionation and identification of some phenolics extracted from evening primrose seed meal. *Czech J. Food Sci.* **2008**, *26*, 58–64. [CrossRef]
56. Ratz-Łyko, A.; Arct, J.; Pytkowska, K.; Majewski, S.; Bregisz, M. Effect of enzymatic hydrolysis on the antioxidant properties of alcoholic extracts of oilseed cakes. *Food Technol. Biotechnol.* **2013**, *51*, 539–546.
57. Wettasinghe, M.; Shahidi, F.; Amarowicz, R. Identification and quantification of low molecular weight phenolic antioxidants in seeds of evening primrose (*Oenothera biennis* L.). *J. Agr. Food Chem.* **2002**, *50*, 1267–1271. [CrossRef]
58. Hadidi, M.; Ibarz, A.; Pouramin, S. Optimization of extraction and deamidation of edible protein from evening primrose (*Oenothera biennis* L.) oil processing by-products and its effect on structural and techno-functional properties. *Food Chem.* **2021**, *334*, 127613. [CrossRef] [PubMed]
59. Kim, H.; Park, S.Y.; Chung, D.K. Effect of the Oral Administration of Common Evening Primrose Sprout (*Oenothera biennis* L.) Extract on Skin Function Improvement in UVB-irradiated Hairless Mice. *Pharmaceuticals* **2021**, *14*, 222. [CrossRef]
60. Kim, T.H.; Shin, H.Y.; Park, S.Y.; Kim, H.; Chung, D.K. Development and Validation of a Method for Determining the Quercetin-3-O-glucuronide and Ellagic Acid Content of Common Evening Primrose (*Oenothera biennis*) by HPLC-UVD. *Molecules* **2021**, *26*, 267. [CrossRef]
61. Farag, M.A.; Reda, A.; Nabil, M.; Elimam, D.M.; Zayed, A. Evening primrose oil: A comprehensive review of its bioactives, extraction, analysis, oil quality, therapeutic merits, and safety. *Food Funct.* **2023**, *14*, 8049–8070. [CrossRef]
62. Kowalewski, Z.; Kowalska, M.; Skrzypezakowa, L. Flavonoids of *Oenothera biennis*. *Diss. Pharm. Pharmacol.* **1968**, *20*, 573–575.
63. Anstett, D.N.; Chen, W.; Johnson, M.T.J. Latitudinal Gradients in Induced and Constitutive Resistance against Herbivores. *J. Chem. Ecol.* **2016**, *42*, 772–781. [CrossRef]
64. Parker, J.D.; Salminen, J.-P.; Agrawal, A.A. Evolutionary Potential of Root Chemical Defense: Genetic Correlations with Shoot Chemistry and Plant Growth. *J. Chem. Ecol.* **2012**, *38*, 992–995. [CrossRef]
65. Van Doormaal, J.J.; Idema, I.G.; Muskiet, F.A.J.; Martini, I.A.; Doorenbos, H. Effects of Short-Term High Dose Intake of Evening Primrose Oil on Plasma and Cellular Fatty Acid Compositions, α-Tocopherol Levels, and Erythropoiesis in Normal and Type 1 (Insulin-Dependent) Diabetic Men. *Diabetologia* **1988**, *31*, 576–584. [CrossRef]
66. Singer, P.; Moritz, V.; Wirth, M.; Berger, I.; Forster, D. Blood Pressure and Serum Lipids from SHR after Diets Supplemented with Evening Primrose, Sunflowerseed or Fish Oil. *Prostaglandins Leukot. Essent. Fat.* **1990**, *40*, 17–20. [CrossRef]
67. Zaitone, S.A.; Moustafa, Y.M.; Mosaad, S.M.; El-Orabi, N.F. Effect of evening primrose oil and ω-3 polyunsaturated fatty acids on the cardiovascular risk of celecoxib in rats. *J. Cardiovasc. Pharmacol.* **2011**, *58*, 72–79. [CrossRef]
68. Uccella, R.; Contini, A.; Sartorio, M. L'azione dell'olio di "evening primrose" sui fattori di rischio cardiovascolare dei diabetici insulino-dipendenti [Action of evening primrose oil on cardiovascular risk factors in insulin-dependent diabetics]. *La Clin. Ter.* **1989**, *129*, 381–388.
69. De La Cruz, J.P.; Martin-Romero, M.; Carmona, J.A.; Villalobos, M.A.; De La Cuesta, F.S. Effect of Evening Primrose Oil On Platelet Aggregation in Rabbits Fed an Atherogenic Diet. *Thromb. Res.* **1997**, *87*, 141–149. [CrossRef] [PubMed]

70. Fan, Y.Y.; Chapkin, R.S.; Ramos, K.S. Dietary lipid source alters murine macrophage/vascular smooth muscle cell interaction in vitro. *J. Nutr.* **1996**, *126*, 2083–2088. [CrossRef] [PubMed]
71. Fan, Y.Y.; Ramos, K.S.; Chapkin, R.S. Dietary gamma-linolenic acid suppresses aortic smooth muscle cell proliferation and modifies atherosclerotic lesions in apolipoprotein E knockout mice. *J. Nutr.* **2001**, *131*, 1675–1681. [CrossRef]
72. Villalobos, M.A.; De La Cruz, J.P.; Martín-Romero, M.; Carmona, J.A.; Smith-Agreda, J.M.; Sánchez de la Cuesta, F. Effect of dietary supplementation with evening primrose oil on vascular thrombogenesis in hyperlipemic rabbits. *Thromb. Haemost.* **1998**, *80*, 696–701. [PubMed]
73. Sadi, A.M.; Toda, T.; Oku, H.; Hokama, S. Dietary effects of corn oil, oleic acid, perilla oil, and evening [corrected] primrose oil on plasma and hepatic lipid level and atherosclerosis in Japanese quail. *Exp. Ani* **1996**, *45*, 55–62. [CrossRef]
74. Mosaad, S.M.; Zaitone, S.A.; Ahmed, A.A.; Abo-Elmatty, D.M.; El-Baz, A.A.; Moustafa, Y.M. Evening primrose oil or forskolin ameliorates celecoxib-enhanced upregulation of tissue factor expression in mice subjected to lipopolysaccharide-induced endotoxemia. *Naunyn Schmiedebergs Arch. Pharmacol.* **2017**, *390*, 483–492. [CrossRef]
75. Andjic, M.; Draginic, N.; Radoman, K.; Jeremic, J.; Turnic, T.N.; Srejovic, I.; Zivkovic, V.; Kovacevic, M.; Bolevich, S.; Jakovljevic, V. Flaxseed and evening primrose oil slightly affect systolic and diastolic function of isolated heart in male but not in female rats. *Int. J. Vitam. Nutr. Res.* **2021**, *91*, 99–107. [CrossRef]
76. Jack, A.M.; Keegan, A.; Cotter, M.A.; Cameron, N.E. Effects of diabetes and evening primrose oil treatment on responses of aorta, corpus cavernosum and mesenteric vasculature in rats. *Life Sci.* **2002**, *71*, 1863–1877. [CrossRef]
77. Stevens, E.J.; Carrington, A.L.; Tomlinson, D.R. Prostacyclin release in experimental diabetes: Effects of evening primrose oil. *Prostaglandins Leukot. Essent. Fat.* **1993**, *49*, 699–706. [CrossRef]
78. Singer, P.; Hoffmann, P.; Beitz, J.; Förster, W.; Wirth, M.; Gödicke, W. Serum triglycerides and HDL cholesterol from SHR after evening primrose oil and other polyunsaturated fats. *Prostaglandins Leukot. Med.* **1986**, *22*, 173–177. [CrossRef] [PubMed]
79. Sugano, M.; Ide, T.; Ishida, T.; Yoshida, K. Hypocholesterolemic effect of gamma-linolenic acid as evening primrose oil in rats. *Ann. Nutr. Metab.* **1986**, *30*, 289–299. [CrossRef] [PubMed]
80. Fragoso, Y.D.; Skinner, E.R. The effect of gammalinolenic acid on the subfractions of plasma high-density lipoprotein of the rabbit. *Biochem. Pharmacol.* **1992**, *44*, 1085–1090. [CrossRef]
81. Mert, H.; İrak, K.; Çibuk, S.; Yıldırım, S.; Mert, N. The effect of evening primrose oil (*Oenothera biennis*) on the level of adiponectin and some biochemical parameters in rats with fructose-induced metabolic syndrome. *Arch. Physiol. Biochem.* **2022**, *128*, 1539–1547. [CrossRef]
82. Zeng, G.; Ju, Y.; Shen, H.; Zhou, N.; Huang, L. Immunopotentiating activities of the purified polysaccharide from evening primrose in H22 tumor-bearing mice. *Int. J. Biol. Macromol.* **2013**, *52*, 280–285. [CrossRef]
83. Lee, J.H.; Kim, M.J.; Woo, K.J.; Hong, J.; Kim, S.H.; Kim, T.J. Evening Primrose Extracts Inhibit PDGF-BB-Induced Vascular Smooth Muscle Cell Proliferation and Migration by Regulating Cell-Cycle-Related Proteins. *Curr. Issues Mol. Biol.* **2022**, *44*, 1928–1940. [CrossRef]
84. Liu, W.; Yin, D.; Li, N.; Hou, X.; Wang, D.; Li, D.; Liu, J. Influence of Environmental Factors on the Active Substance Production and Antioxidant Activity in *Potentilla fruticosa* L. and Its Quality Assessment. *Sci. Rep.* **2016**, *6*, 28591. [CrossRef]
85. Chelghoum, M.; Guenane, H.; Tahri, D.; Laggoun, I.; Marfoua, F.Z.; Rahmani, F.Z.; Khenifer, F.; Yousfi, M. Influence of Altitude, Precipitation, and Temperature Factors on the Phytoconstituents, Antioxidant, and α-Amylase Inhibitory Activities of *Pistacia atlantica*. *J. Food Meas. Charact.* **2021**, *15*, 4411–4425. [CrossRef]
86. Wulff, A.; Anttonen, S.; Pellinen, R.; Savonen, E.M.; Sutinen, M.L.; Heller, W.; Sandermann, H.J.; Kangasjärvi, J. Birch (*Betula pendula* Roth.) responses to high UV-B radiation. *Boreal. Environ. Res.* **1999**, *4*, 77–88.
87. Dong, J.; Ma, X.; Wei, Q.; Peng, S.; Zhang, S. Effects of growing location on the contents of secondary metabolites in the leaves of four selected superior clones of *Eucommia ulmoides*. *Ind. Crops Prod.* **2011**, *3*, 1607–1614. [CrossRef]
88. Suyal, R.; Rawat, S.; Rawal, R.S.; Bhatt, I.D. Variability in morphology, phytochemicals, and antioxidants in *Polygonatum verticillatum* (L.) All. populations under different altitudes and habitat conditions in Western Himalaya, India. *Environ. Monit. Assess.* **2020**, *191* (Suppl. 3), 783. [CrossRef] [PubMed]
89. ANM [Administrația Națională de Meteorologie]. *Clima României*; Editura Academiei Române: Bucharest, Romania, 2008.
90. Păcurar, F. *Specii Indicator Pentru Evaluarea și Elaborarea Managementului Sistemelor de Pajiști cu Înaltă Valoare Naturală–HNV*; Casa Cărții de Știință: Cluj-Napoca, Romania, 2020; pp. 16–18.
91. Ielciu, I.; Filip, G.A.; Oniga, I.; Olah, N.-K.; Bâldea, I.; Olteanu, D.; Burtescu, R.F.; Turcuș, V.; Sevastre-Berghian, A.C.; Benedec, D.; et al. Oxidative Stress and DNA Lesion Reduction of a Polyphenolic Enriched Extract of *Thymus marschallianus* Willd. in Endothelial Vascular Cells Exposed to Hyperglycemia. *Plants* **2021**, *10*, 2810. [CrossRef] [PubMed]
92. Mihali, C.V.; Petrescu, C.M.; Ciolacu-Ladasiu, C.F.; Mathe, E.; Popescu, C.; Bota, V.; Mizeranschi, A.E.; Ilie, D.E.; Neamț, R.I.; Turcus, V. Assessing Phenotypic Variability in Some Eastern European Insular Populations of the Climatic Relict *Ilex aquifolium* L. *Plants* **2022**, *11*, 2022. [CrossRef]

93. Bota, V.B.; Neamtu, A.-A.; Olah, N.-K.; Chișe, E.; Burtescu, R.F.; Pripon Furtuna, F.R.; Nicula, A.-S.; Neamtu, C.; Maghiar, A.-M.; Ivănescu, L.-C.; et al. A Comparative Analysis of the Anatomy, Phenolic Profile, and Antioxidant Capacity of *Tussilago farfara* L. Vegetative Organs. *Plants* **2022**, *11*, 1663. [CrossRef]
94. Chytrý, M.; Tichý, L.; Dřevojan, P.; Sádlo, J.; Zelený, D. Ellenberg-type indicator values for the Czech flora. *Preslia* **2018**, *90*, 83–103. [CrossRef]
95. Kosakowska, O.; Węglarz, Z.; Przybył, J.L. Chemical and genetic diversity of evening primrose (*Oenothera biennis* L.) occurring in the eastern area of Poland. *Acta Hortic.* **2008**, *765*, 151–156. [CrossRef]
96. Kim, T.H.; Kim, W.J.; Park, S.Y.; Kim, H.; Chung, D.K. In Vitro Anti-Wrinkle and Skin-Moisturizing Effects of Evening Primrose (*Oenothera biennis*) Sprout and Identification of Its Active Components. *Processes* **2021**, *9*, 145. [CrossRef]
97. Anstett, D.N.; Ahern, J.R.; Johnson, M.T.J.; Salminen, J.P. Testing for latitudinal gradients in defense at the macroevolutionary scale. *Evolution* **2018**, *72*, 2129–2143. [CrossRef]
98. Yan, J.; Yu, L.; Xu, S.; Gu, W.; Zhu, W. Apigenin accumulation and expression analysis of apigenin biosynthesis relative genes in celery. *Sci. Hortic.* **2014**, *165*, 218–224. [CrossRef]
99. Šamec, D.; Karalija, E.; Šola, I.; Vujčić Bok, V.; Salopek-Sondi, B. The Role of Polyphenols in Abiotic Stress Response: The Influence of Molecular Structure. *Plants* **2021**, *10*, 118. [CrossRef]
100. Smith, G.J.; Markham, K.R. Tautomerism of flavonol glucosides: Relevance to plant UV protection and flower colour. *J. Photochem. Photobiol. A Chem.* **1988**, *118*, 99–105. [CrossRef]
101. Zhang, L.; Xie, Q.; Li, X. Esculetin: A review of its pharmacology and pharmacokinetics. *Phytother. Res.* **2021**, *36*, 279–298. [CrossRef] [PubMed]
102. Tattini, M.; Di Ferdinando, M.; Brunetti, C.; Goti, A.; Pollastri, S.; Bellasio, C.; Giordano, C.; Fini, A.; Agati, G. Esculetin and esculin (esculetin 6-O-glucoside) occur as inclusions and are differentially distributed in the vacuole of palisade cells in Fraxinus ornus leaves: A fluorescence microscopy analysis. *J. Photochem. Photobiol. B* **2014**, *140*, 28–35. [CrossRef]
103. Wang, Y.; Cui, Z.; Li, Q.; Zhang, S.; Li, Y.; Li, X.; Kong, L.; Luo, J. The parallel biosynthesis routes of hyperoside from naringenin in *Hypericum monogynum*. *Hortic. Res.* **2023**, *10*, uhad166. [CrossRef]
104. Hara, M.; Furukawa, J.; Sato, A.; Mizoguchi, T.; Miura, K. Abiotic Stress and Role of Salicylic Acid in Plants. In *Abiotic Stress Responses in Plants*; Ahmad, P., Prasad, M., Eds.; Springer: New York, NY, USA, 2012. [CrossRef]
105. Ozfidan-Konakci, C.; Yildiztugay, E.; Alp, F.N.; Kucukoduk, M.; Turkan, I. Naringenin induces tolerance to salt/osmotic stress through the regulation of nitrogen metabolism, cellular redox and ROS scavenging capacity in bean plants. *Plant Physiol. Biochem.* **2020**, *157*, 264–275. [CrossRef]
106. Legea Plantelor Medicinale și Aromatice, nr. 491/2003. Available online: https://lege5.ro/Gratuit/gq3denbx/legea-nr-491-2003-privind-plantele-medicinale-si-aromatice-precum-si-produsele-stupului (accessed on 25 June 2020).
107. Ordonața de Urgență 57/2007 Privind Regimul Ariilor Naturale Protejate, Conservarea Habitatelor Naturale, a Florei și Faunei Sălbatice. Monitorul Oficial al României. 2007. Available online: https://legislatie.just.ro/Public/DetaliiDocument/83289 (accessed on 25 June 2020).
108. Directive 92/43/EEC. Available online: https://eur-lex.europa.eu/legal-content/en/NIM/?uri=CELEX:31992L0043 (accessed on 25 June 2020).
109. Pirnă, I.; Voicu, E.; Radu, M.; Ciupercă, R.; Muscalu, A. Ghid. Tehnologii de Recoltare a Plantelor Medicinale Și Aromatice Din Zona Călărași-Silistra. Institutul Național de Cercetare-Dezvoltare pentru Științe Biologice. 2012. Available online: https://proiecte.incdsb.ro/medplanet/doc/Ghid%20tehnologii%20de%20recoltare%20RO.pdf (accessed on 25 June 2020).
110. European Directorate for the Quality of Medicines & HealthCare (EDQM). *European Pharmacopoeia*; EDQM IX; Council of Europe: Strasbourg, France, 2021.
111. *Homöopathische Arzneibuch*; HAB 2021; Deutscher Apotheker Verlag: Stuttgart, Germany, 2021.
112. Editura Medicala. *Romanian Pharmacopoeia*; RPh. X; Editura Medicala: Bucharest, Romania, 1993; pp. 1035–1041.
113. Meteomanz Database. Available online: http://www.meteomanz.com/ (accessed on 10 January 2022).
114. QGIS Development Team. QGIS Geographic Information System. Version 3.40 "Bratislava"; Open Source Geospatial Foundation Project: 2024. Available online: https://qgis.org (accessed on 25 September 2025).
115. Philip, G.M.; Watson, D.F. A Precise Method for Determining Contoured Surfaces. *APEA J.* **1982**, *22*, 205–212. [CrossRef]
116. Watson, D.F.; Philip, G.M. A Refinement of Inverse Distance Weighted Interpolation. *Geoprocessing* **1985**, *2*, 315–327.
117. Goodchild, M.F. Citizens as sensors: The world of volunteered geography. *GeoJournal* **2007**, *69*, 211–221. [CrossRef]
118. Longley, P.A.; Goodchild, M.F.; Maguire, D.J.; Rhind, D.W. *Geographic Information Science and Systems*, 5th ed.; Wiley: Hoboken, NJ, USA, 2021.
119. Crampton, J.W. *Mapping: A Critical Introduction to Cartography and GIS*; Wiley-Blackwell: Chichester, UK, 2010.
120. Nistor, M.M.; Nicula, A.-S.; Cervi, F.; Man, T.-C.; Irimuș, I.-A.; Surdu, I. Groundwater vulnerability GIS models in the Carpathian Mountains under climate and land cover changes. *Appl. Ecol. Environ. Res.* **2018**, *16*, 5095–5116. [CrossRef]

121. Nicula, A.-S.; Stoica, M.-S.; Bîrsănuc, E.-M.; Man, T.-C. Why do Romanians take to the streets? A spatial analysis of Romania' 2016–2017 protests. *Rom. J. Political Sci.* **2019**, *19*, 201–222.
122. Nicula, A.-S.; Boțan, C.N.; Gligor, V.; Cociș, E.-A. Celebrating the Great Union through smart digital solutions: Lessons from Alb Iulia, Romania. *J. Urban Hist.* **2022**, *48*, 425–443. [CrossRef]
123. Nicula, A.-S.; Stoica, M.-S.; Ilovan, O.-R. The cultural-historical and political spheres of influence of Alba Iulia. *Transylv. Rev.* **201** *26* (Suppl. 2), 299–315.
124. Bîrsănuc, E.-M.; Cociș, E.-A.; Gligor, V.; Man, T.-C.; Nicula, A.-S.; Stoica, M.-S. Performing democracy: An analysis of church-base electoral capital in Romania. *Transylv. Rev.* **2020**, *29* (Suppl. 2), 291–309. [CrossRef]
125. Nicula, A.-S.; Nistor, M.M. Application of GIS Technology for Tourism Flow Modelling in the United Kingdom. *Geogr. Tech.* **202** *16*, 1–12. [CrossRef]
126. Nistor, M.M.; Nicula, A.-S.; Dezsi, S.; Petrea, D.; Kamarajugedda, S.A.; Carebia, I.-A. GIS-based kernel analysis for tourism flow mapping. *J. Settl. Spat. Plan.* **2020**, *11*, 137–145. [CrossRef]
127. Marosi, Z.; Adorean, E.-C.; Ilovan, O.-R.; Gligor, V.; Voicu, C.-G.; Nicula, A.-S.; Dulamă, M.-E. Living the urban cultural landscape in the city centre of Cluj Napoca/Kolozsvár/Klausenburg, Romania. *MÖGG* **2019**, *161*, 117–160. [CrossRef]
128. Nistor, M.M.; Nicula, A.-S.; Haidu, I.; Surdu, I.; Carebia, I.-A.; Petrea, D. GIS Integration Model of Metropolitan Area Sustainability Index (MASI). The Case of Paris Metropolitan Area. *J. Settl. Spat. Plan.* **2019**, *10*, 39–48. [CrossRef]

molecules

MDPI

Article

Microwave-Assisted Extraction of Phenolic Compounds from Cocoa Pod Husk: Process Optimization and Impact of Drying Temperature on Bioactive Recovery

Pablo Gomez [1], Cristhopher Reyes [1] and Jorge G. Figueroa [2,*]

[1] Carrera de Ingeniería Química, Facultad de Ciencias Exactas y Naturales, Universidad Técnica Particular de Loja, París s/n y Praga, Loja 110107, Ecuador; psgomez1@utpl.edu.ec (P.G.); careyes19@utpl.edu.ec (C.R.)
[2] Departamento de Química, Facultad de Ciencias Exactas y Naturales, Universidad Técnica Particular de Loja, París s/n y Praga, Loja 110107, Ecuador
* Correspondence: jgfigueroa@utpl.edu.ec; Tel.: +593-7-3701444

Abstract

Cocoa pod husk (CPH), the principal by-product of cocoa processing, represents an abundant and underutilized source of bioactive phenolics with potential applications in the food and nutraceutical sectors. This study optimized the extraction of catechin, epicatechin, procyanidin B2, and clovamide from CPH (CCN-51 variety) using microwave-assisted extraction (MAE) and evaluated the influence of drying temperature on their retention. A Box–Behnken design within a response surface methodology framework was employed to evaluate the effects of ethanol concentration (0–100%), extraction temperature (50–150 °C), and extraction time (15–60 min) on compound recovery. The phenolic profile was characterized by high-performance liquid chromatography with diode-array detection and electrospray ionization ion trap tandem mass spectrometry. Optimal MAE conditions of 51% ethanol, 104 °C, and 38 min yielded maximum concentrations of clovamide, procyanidin B2, and epicatechin of 3440, 908, and 445 mg/kg dry matter of cocoa pod husk, respectively. Drying studies demonstrated that moderate hot-air temperatures (40–50 °C) preserved the highest phenolic levels. These results underscore the importance of optimizing both extraction and drying conditions to enhance the recovery of phenolic compounds from cocoa processing residues, supporting their potential valorization as antioxidant-rich functional ingredients.

Keywords: cocoa pod husk; microwave-assisted extraction; catechin; epicatechin; procyanidin B2; clovamide; response surface methodology; drying temperature; HPLC-DAD-ESI-IT-MS/MS

1. Introduction

Theobroma cacao L., a member of the Malvaceae family, is a perennial tropical tree native to Central and South America [1]. The principal producers are Ivory Coast, Ghana, Indonesia, Ecuador, and Brazil [2]. Its seeds, known as cocoa beans, are valued for their distinctive flavor and nutritional properties, contributing significantly to the global economy [3]. In 2023, approximately 5.6 million metric tons of cocoa beans were processed worldwide [2]. This large-scale production generates considerable amounts of by-products, with approximately two kilograms produced for every kilogram of beans [4]. These by-products consist mainly of cocoa pod husk (CPH), mucilage, and bean shell, with CPH accounting for around 86% of the total biomass. Although traditionally underused, CPH contains high levels of phenolic bioactive compounds with documented antioxidant, anti-inflammatory, and antidiabetic properties [4–6]. These include protocatechuic acid, *p*-hydroxybenzoic acid, salicylic acid, kaempferol, linarin, resveratrol, apigenin, and luteolin [7,8], which confer significant potential for its valorization into functional ingredients. Nevertheless, due to its high moisture content (approximately 90%), CPH must first be dehydrated to prevent microbial spoilage and ensure material stability [3,5].

In this context, several drying techniques, including sun drying, freeze drying, vacuum drying, and forced convection drying, have been applied to CPH [4,6,9]. Sun drying, although inexpensive and widely accessible, is highly dependent on climatic conditions and presents an elevated risk of microbial contamination [4]. Freeze-drying ensures outstanding preservation of thermolabile compounds [10]. Nevertheless, its large-scale implementation is constrained by substantial capital requirements and high operational energy demand [5,8]. Vacuum drying reduces oxidative degradation but may result in heterogeneous moisture removal within the biomass [10]. Forced convection drying stands out among the available techniques because it allows precise control over temperature and airflow, which in turn ensures consistent moisture removal across the biomass [3,9]. Furthermore, forced convection drying is adaptable to different scales of operation, from laboratory to industrial levels, and offers shorter processing times while maintaining reproducibility and product quality [10]. These advantages make it a particularly attractive option for the stabilization of CPH, despite its significant energy requirements and dedicated infrastructure [10]. Nonetheless, most bioactive compounds are sensitive and prone to degradation when exposed to drying conditions such as high temperatures and oxygen-rich environments [11]. Therefore, it is crucial to evaluate the influence of drying temperature on their stability in order to establish conditions that safeguard both chemical integrity and functional properties [4].

Once stabilized through drying, efficient extraction becomes the next critical step, requiring methods that maximize yields while preserving compound integrity [12]. In this context, advanced techniques such as microwave-assisted extraction (MAE), pressurized liquid extraction, supercritical fluid extraction, and ultrasound-assisted extraction have been widely studied. These extraction methods are considered environmentally friendly, as they reduce solvent consumption and processing time. Nevertheless, each presents inherent technical and operational constraints. Pressurized liquid extraction and supercritical fluid extraction require high-pressure systems and high energy input, which restrict their industrial feasibility [13]. Ultrasound-assisted extraction may be effective for some matrices, but it often shows low efficiency for certain phenolic structures and can induce free radical formation that compromises stability [14]. MAE provides high efficiency and selectivity for phenolics, although it may cause non-uniform heating in heterogeneous matrices and increase the risk of thermal degradation of thermolabile compounds [15].

Among these alternatives, MAE stands out for its ability to use microwave energy to disrupt cellular matrices and promote the rapid release of intracellular compounds, typically with polar solvents that efficiently absorb microwave energy and enhance mass transfer [16]. MAE is also inherently adaptable to industrial-scale operations through appropriate reactor design and process optimization [17], making it a promising technology for large-scale phenolic recovery while preserving compound integrity. Given the potential and limitations of MAE, its successful application requires careful optimization of operational parameters to maximize phenolic recovery while preventing thermal degradation. In this context, Response Surface Methodology (RSM) has proven useful in cocoa matrices for defining suitable MAE conditions and improving extraction efficiency [18,19].

Although previous studies have demonstrated the potential of CPH as a rich source of phenolic compounds, most research has addressed extraction optimization or drying effects separately and has relied primarily on bulk phenolic measurements or qualitative identification of individual compounds without detailed molecular profiling. Rahayu, et al. [20], Van, et al. [21], and Mashuni, Hamid, Muzuni, Kadidae, Jahiding, Ahmad, and Saputra [5] evaluated MAE using ethanol–water mixtures, but their analyses were restricted to total phenolics and basic identification of catechin, without compound-specific quantification or multivariate statistical optimization. Other works have emphasized CPH valorization using alternative extraction methods such as sonotrode-assisted extraction or optimization of drying strategies such as vacuum or forced convection drying [6,8]. Nonetheless, none of these studies investigated how drying-induced transformations impact the recovery and structural stability of specific thermolabile compounds, including procyanidin B2 and clovamide, both of which are prone to enzymatic and thermal degradation. Furthermore, Cádiz-Gurrea, Fernández-Ochoa, Leyva-Jiménez, Guerrero-Muñoz, Villegas-Aguilar, Pimentel-Moral, Ramos-Escudero, and Segura-Carretero [3] provided only qualitative chemical identification without quantification or integration with process optimization frameworks. Therefore, this study aimed to optimize the extraction of phenolic compounds from CPH using MAE with generally recognized as safe (GRAS) solvents and to assess the effect of drying temperature on the recovery and stability of key bioactives, namely catechin, epicatechin, procyanidin B2, and clovamide.

2. Results and Discussion

2.1. Identification of Phenolic Compounds in Cocoa Pod Husk

Catechin and epicatechin, eluting at RT = 20.2 min and RT = 23.7 min, respectively, both exhibited $[M-H]^-$ ions at m/z 289.2 and identical characteristic fragments at m/z 165, 179, 245, and 271 (Table 1). These fragment ions result from typical retro-Diels–Alder cleavage and heterocyclic ring fission processes associated with the C-ring of flavan-3-ols, consistent with previously reported fragmentation profiles for catechin and epicatechin [9].

Table 1. Mass spectrometric characteristics of phenolic compounds identified in cocoa pod husk by high-performance liquid chromatography with diode-array detection and electrospray ionization ion-trap tandem mass spectrometry.

Compound	RT (min)	m/z	MS/MS Fragments
Catechin	20.2	289.2	165, 179, 245, 271
Procyanidin B2	20.9	577.3	205, 289, 425, 559
Epicatechin	23.7	289.2	165, 179, 245, 271
Clovamide	28.4	358.2	178, 222, 314

Procyanidin B2, detected at RT = 20.9 min, was identified by its $[M-H]^-$ ion at m/z 577.3 and diagnostic fragments at m/z 205, 289, 425, and 559 (Table 1). These signals correspond to sequential quinone methide and heterocyclic ring cleavages, including the loss of catechin/epicatechin monomer units, confirming its structure as a B-type procyanidin dimer [22].

Clovamide, eluting at RT = 28.4 min (Table 1), was tentatively identified as $[M-H]^-$ at m/z 358.2. Its MS/MS fragments at m/z 178, 222, and 314 reflect characteristic cleavages of its caffeic acid and L-DOPA-derived substructures, as previously described for phenolic amides in cocoa-related matrices [8].

2.2. Optimization of Phenolic Compound Extraction by MAE

Extraction is a critical step in maximizing the recovery of target compounds from plant matrices [8]. Several factors influence the efficiency of the extraction process, including the extraction method, solvent type, temperature, extraction time, and solvent-to-sample ratio, as well as sample-related characteristics such as fat content, moisture level, and particle size [23]. In this study, ethanol concentration, extraction temperature, and extraction time were selected as independent variables, while solvent volume, sample mass, and particle size were kept constant.

MAE was chosen due to its advantages over conventional extraction methods. MAE offers shorter extraction times, reduced solvent consumption, higher extraction efficiency, and lower processing costs [15].

With regard to the solvent system, ethanol and water were selected as extraction media, both of which are classified as GRAS and therefore suitable for use in the food and nutraceutical industries [13]. Moreover, the combination of ethanol and water has been shown to enhance the extraction efficiency of phenolic acids and flavonoids [13].

Temperature is another key factor, as elevated temperatures promote the rupture of plant cell structures, facilitating the release of phenolic compounds and improving extraction yields [24]. Nevertheless, excessively high temperatures should be avoided due to the thermal sensitivity of polyphenols, which are prone to degradation under such conditions [12]. Accordingly, extraction temperature was evaluated in the range of 50 to 150 °C.

Extraction time also plays a critical role in MAE, a technique known for its ability to heat the sample uniformly both internally and externally without thermal gradients. As a result, extraction can be achieved in shorter durations [12]. For this reason, extraction time was evaluated within the range of 15 to 60 min.

The response variables selected for optimization were the concentrations of procyanidin B2, catechin, epicatechin, and clovamide. Notably, procyanidins, catechin, and epicatechin are flavonoids belonging to the polyphenol family. These compounds possess antioxidant capacities 20 to 50 times higher than those of vitamins C and E, respectively [25]. In addition to their antioxidant potential, they have demonstrated various health-promoting effects, including glycemic and insulin regulation, neuroprotective properties, anti-tumor activity, and anti-inflammatory effects [26].

Regarding clovamide, this compound is of particular interest due to its notable antioxidant activity. It has been shown to exhibit free radical scavenging properties comparable to those of rosmarinic acid and caffeic acid and to be more effective than other well-known antioxidants such as epicatechin [27]. These findings highlight clovamide as a promising bioactive compound for pharmaceutical and food-related research, with potential applications as a nutraceutical ingredient.

The BBD, integrated within the framework of RSM, was employed to optimize the extraction of phenolic compounds from cocoa by-products and to evaluate the combined effects of ethanol concentration, extraction temperature, and extraction time. This approach enables efficient analysis of multiple variables while minimizing the number of experimental runs, time, and material consumption. To assess the adequacy of the mathematical model, an analysis of variance (ANOVA) was performed, followed by Fisher's statistical test to determine the significance of each independent variable.

The adequacy and reliability of the statistical models obtained for the extraction process were evaluated through ANOVA. The resulting quadratic models exhibited R^2 ranging from 73% to 91%, indicating excellent agreement between experimental data and predicted values and confirming the suitability of these models for predicting extraction behavior [28].

Ethanol concentration showed a pronounced quadratic effect ($p < 0.05$) on the extraction yield of all four compounds, clearly demonstrating optimal points within the tested range (0–100%). This observation highlights the critical role that ethanol-water mixtures play in phenolic compound extraction. Specifically, intermediate ethanol concentrations (~50%) consistently produced the highest yields, which can be attributed to the synergistic solvation properties of binary solvent systems. Such mixtures improve the polarity balance of the extraction medium, facilitating the solubilization of both hydrophilic and moderately lipophilic phenolics, enhancing mass transfer, and helping to stabilize thermolabile compounds during extraction [29].

The response surface plots (Figure 1) and regression models demonstrated a significant negative quadratic effect ($p < 0.05$) of ethanol concentration on the recovery of catechin, epicatechin, procyanidin B2, and clovamide, with coefficients of −0.00949, −0.0668, −0.1979, and −0.654, respectively. These results indicate that extraction efficiency peaked at intermediate ethanol–water ratios and declined toward both extremes of the tested range, a behavior consistent with the influence of solvent polarity on phenolic extraction.

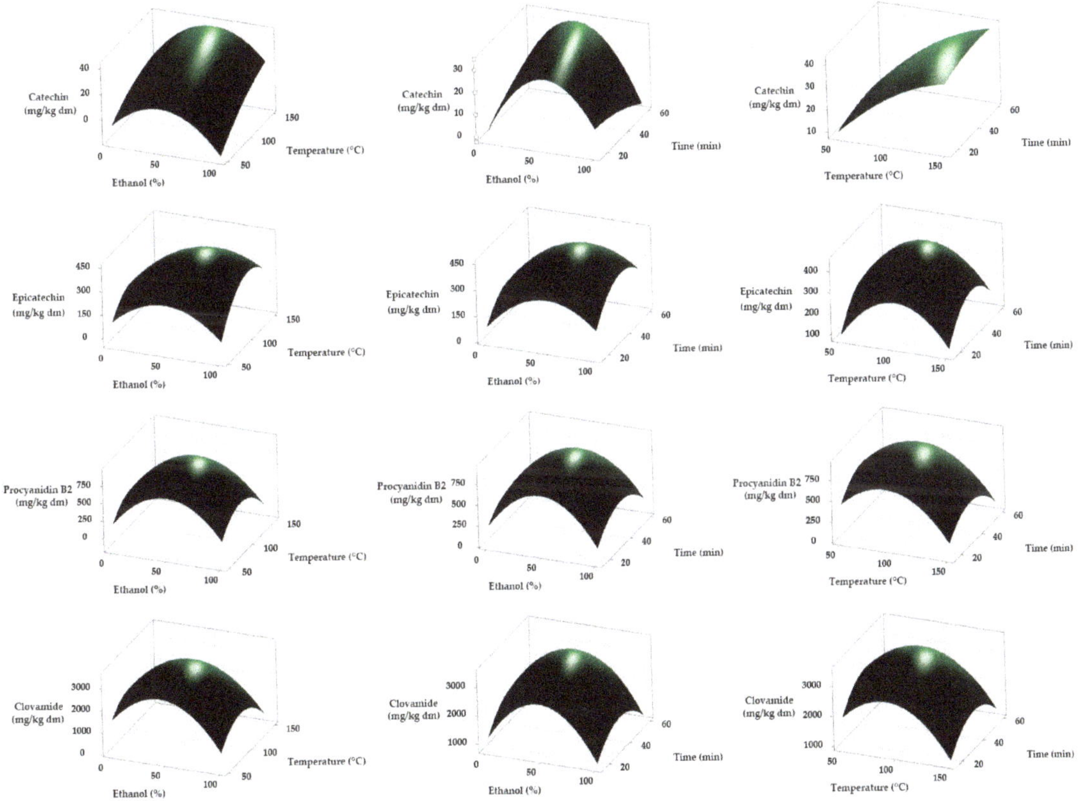

Figure 1. Response surface plots showing the effects of ethanol concentration (%), extraction temperature (°C), and extraction time (min) on the concentration of phenolic compounds from cocoa pod husk: catechin, epicatechin, procyanidin B2, and clovamide.

Hydroethanolic mixtures at intermediate proportions are more effective than single solvents because their balanced polarity improves the solubility of compounds with different polarities, while the aqueous fraction facilitates desorption of analytes from the plant matrix. This combination of solvent effects has been explicitly reported for MAE and related techniques [12,26].

This interpretation is consistent with findings from other MAE studies in various plant matrices, where ethanol–water mixtures around 45–60% yielded higher total phenolic contents than either pure water or pure ethanol [19,29]. The agreement with published data from other polyphenol-rich sources supports the conclusion that intermediate ethanol concentrations provide an optimal polarity balance, improving extraction yield while preserving compound integrity.

Extraction temperature significantly influenced the recovery of the target polyphenols, although the nature of the effect differed among compounds. For epicatechin, procyanidin B2, and clovamide, regression models revealed a significant quadratic term for temperature ($p = 0.003$, $p = 0.005$, and $p = 0.019$, respectively), indicating that maximum extraction efficiency occurred at intermediate values. Within this range, MAE promotes efficient solvent penetration and cell disruption while limiting degradation, as the rapid dielectric heating of the intracellular water causes cell wall rupture and accelerates the release of phenolic compounds without prolonged exposure to thermal stress [30]. Beyond this optimum, concentrations declined markedly, consistent with the negative quadratic coefficients and the known thermolability of these compounds.

Procyanidin B2, a dimeric flavan-3-ol composed of two epicatechin units linked by an interflavan C4–C8 bond, is particularly susceptible to thermal depolymerization [25]. Moderate heating may liberate monomeric units, but further heating leads to oxidative degradation, a pathway more pronounced than in the monomeric catechin. Catechin showed no significant quadratic temperature effect ($p = 0.572$) but exhibited a positive linear relationship ($p = 0.039$), suggesting that the apparent steady increase in catechin concentration within the tested range may be partially explained by the thermal depolymerization of procyanidin B2, which releases monomeric epicatechin units that can undergo epimerization to catechin before significant degradation occurs. Epicatechin, despite also being a monomer, displayed a pronounced quadratic response due to its higher susceptibility to epimerization, oxidative cleavage, and condensation reactions under heat stress [11]. Catechin and epicatechin, although structural isomers, showed distinct thermal behaviors under MAE in this study. Catechin exhibited a positive trend, consistent with formation from procyanidin depolymerization and epimerization of epicatechin, whereas epicatechin peaked at intermediate temperatures before declining due to greater reactivity. Liazid, et al. [31] reported a similar divergence: catechin remained stable up to 100 °C and increased further at 125 °C and 150 °C, likely due to liberation from oligomers, while epicatechin stayed stable up to 100 °C but did not show a comparable increase at higher temperatures, reflecting differences in reactivity despite structural similarity.

Thermal degradation of epicatechin, procyanidin B2, and clovamide involves oxidative cleavage, depolymerization, and structural rearrangements that reduce the concentration of individual compounds in the extract. Comparable effects have been reported in polyphenol-rich matrices such as grape seed and cocoa bean extracts, where excessive heating markedly reduced total procyanidin content [19,31,32]. These findings highlight the importance of optimizing MAE temperature conditions according to the thermal stability and structural characteristics of each phenolic compound to maximize yield while preserving their integrity.

The response surface plots (Figure 1) and regression analyses showed that the influence of extraction time under MAE conditions varied markedly among the phenolic compounds analyzed. For catechin and clovamide, neither the linear nor the quadratic terms for time were statistically significant ($p > 0.07$), indicating that extraction equilibrium was reached rapidly. This suggests that prolonged extraction times offer no measurable advantage for their recovery. In the case of clovamide, the quadratic model was significant overall ($p = 0.044$) due to pronounced curvature effects of solvent composition and temperature, whereas time contributed minimally to model performance, with the quadratic effect of time showing only marginal significance ($p = 0.071$). This trend aligns with the fundamental principles governing MAE, where the ability to heat the sample matrix both internally and externally without a thermal gradient accelerates cell disruption and solute release, thereby achieving maximum yields in the first minutes of extraction [15,16].

In contrast, epicatechin and procyanidin B2 exhibited statistically significant quadratic effects of time ($p = 0.008$ and $p = 0.024$, respectively), resulting in a parabolic response within the tested range (20–60 min). Maximum yields were obtained at intermediate extraction times, while both shorter and longer durations produced lower recoveries. This pattern reflects a balance between rapid solubilization facilitated by dielectric heating and subsequent degradation, epimerization, or reduced mass transfer efficiency when extraction is excessively prolonged [12]. Comparable trends have been reported for MAE of phenolics from other plant matrices. For instance, in *Agaricus blazei* Murrill, extraction yields of several phenolic compounds reached a maximum at intermediate times and subsequently declined, a behavior attributed to thermal degradation and structural transformations during prolonged microwave exposure [29].

The differences in time dependence are likely related to molecular structure and stability. Procyanidin B2 contains multiple hydroxyl groups that are more susceptible to oxidative or thermal degradation [25]. Being a dimeric flavanol, it may also undergo depolymerization under extended microwave exposure. In contrast, clovamide possesses an amide linkage and conjugated aromatic system that confer greater resistance to microwave-induced dielectric and thermal stress.

Overall, the statistical models confirmed that extraction time was a secondary factor compared with solvent composition and temperature for all analytes except epicatechin and procyanidin B2. For catechin and clovamide, short extraction times are sufficient, while intermediate durations are optimal for epicatechin and procyanidin B2 to balance rapid release with minimal degradation. These results support the widely recognized advantage of MAE in achieving high phenolic recoveries in significantly shorter times than conventional methods, driven by its dielectric heating mechanism and uniform energy distribution within the extraction matrix [19,31].

The composite desirability and individual desirability profiles (Figure 2) identified the optimal MAE conditions as 51% ethanol, 104 °C, and 38 min, yielding a composite desirability of 0.837. Clovamide and procyanidin B2 achieved the highest desirabilities (0.961 and 0.919, respectively), corresponding to yields of 3440 mg/kg dm and 908 mg/kg dm. Epicatechin exhibited a desirability of 0.954 with a yield of 445 mg/kg dm, aligning with findings by Maldonado and Figueroa [19], who showed that epicatechins remain stable and well-recovered under moderate-temperature MAE with ethanol-water mixtures and degrade only at excessive heat. In contrast, catechin showed the lowest desirability (0.583) and yield of 33.0 mg/kg dm.

This outcome likely reflects its lower initial concentration in CPH and its sensitivity to extraction temperature. Notably, catechin levels increased at higher temperatures, probably as a result of the thermal depolymerization of higher procyanidin oligomers (dimers, trimers, and tetramers) into monomeric catechin units.

This finding indicates that although elevated temperatures may indirectly promote catechin recovery, they introduce challenges in defining the overall optimal extraction conditions.

The optimized MAE parameters were subsequently employed to assess the effect of drying temperature on the phenolic composition of CPH.

2.3. *Effect of Drying Temperature on Phenolic Content*

Drying temperature exerted a pronounced influence on the retention of catechin, epicatechin, procyanidin B2, and clovamide in CPH (Figure 3). Maximum phenolic concentrations were consistently observed under moderate hot-air drying (40–50 °C), with statistically significant differences among temperatures for all analytes (Tukey's test, $p < 0.05$).

The combined concentration of the four quantified phenolics reached a maximum of 5099 mg/kg dm at 50 °C. In contrast, the lowest totals were observed at

30 °C (1235 mg/kg dm) and 70 °C (2255 mg/kg dm), representing 75.8% and 55.8% reductions, respectively, compared with the maximum. Moderate drying temperatures inactivate degradative enzymes such as polyphenol oxidase and peroxidase [33]. This temperature range also improves extractability by disrupting cell wall integrity and weakening the polysaccharide–protein network, which increases solvent access and facilitates phenolic release [34]. At ≥ 60–70 °C, non-enzymatic oxidative and thermal reactions increase, causing the loss of flavan-3-ols and hydroxycinnamic acid amides [35]. At 30 °C, prolonged moisture retention allows enzymatic oxidation to proceed [1]. These trends agree with reports on CPH and other cocoa tissues dried with hot air or microwaves, where intermediate temperatures favor phenolic retention [9].

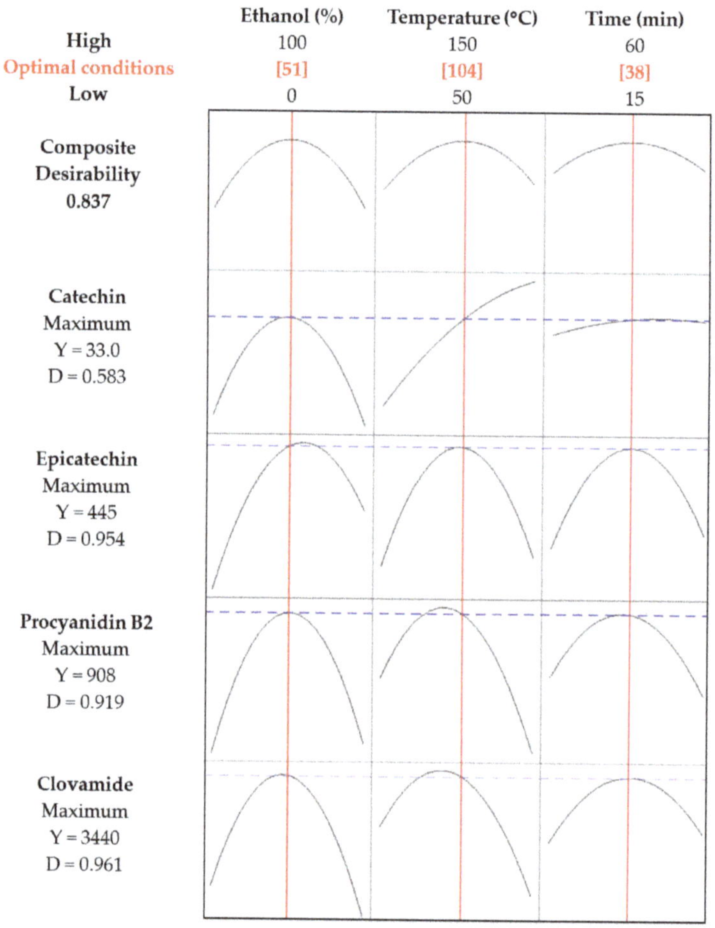

Figure 2. Response optimizer plot showing the desirability-based optimization of ethanol concentration (%), extraction temperature (°C), and extraction time (min) for the microwave-assisted extraction of catechin, epicatechin, procyanidin B2, and clovamide from cocoa pod husk. The red vertical lines indicate the optimal conditions (51% ethanol, 104 °C, 38 min) for maximizing the individual responses and the overall composite desirability (0.837). For each compound, Y represents the predicted maximum concentration achieved at the optimal conditions, while D denotes the individual desirability value (ranging from 0 to 1), indicating how closely the predicted concentration meets the optimization target, with 1 representing the ideal value.

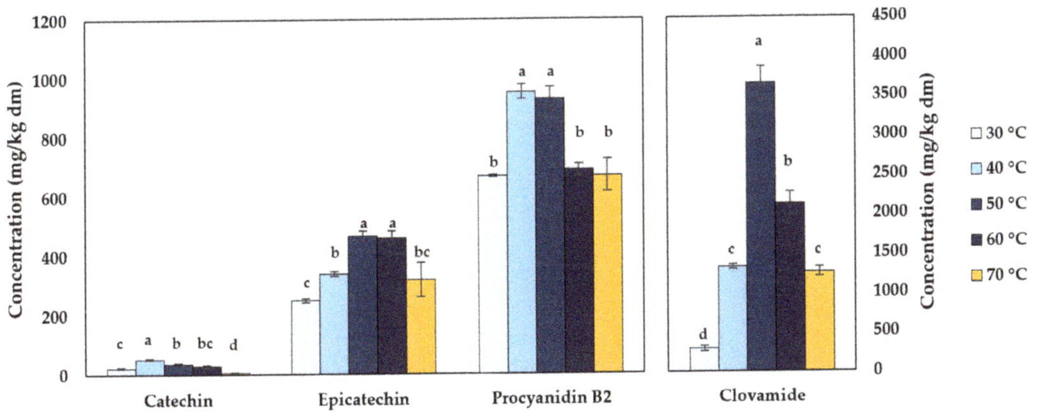

Figure 3. Effect of drying temperature on the concentration of phenolic compounds (catechin, epicatechin, procyanidin B2, and clovamide) in cocoa pod husk. Different letters above bars indicate significant differences among temperatures for each compound (Tukey's test, $p < 0.05$).

Clovamide was the most abundant phenolic, peaking at 3665 ± 184 mg/kg dm at 50 °C. Its concentration increased markedly from 293 ± 34 mg/kg dm at 30 °C to 1325 ± 31 mg/kg dm at 40 °C, before reaching its maximum at 50 °C. Beyond this point, levels dropped significantly to 2126 ± 140 mg/kg dm at 60 °C and 1261 ± 61 mg/kg dm at 70 °C. This biphasic behavior reflects clovamide's dual susceptibility to degradation. At lower temperatures, polyphenol oxidase-mediated oxidation of its caffeic acid moiety likely drives quinone formation and polymerization [36]. At higher temperatures, non-enzymatic decarboxylation and free-radical oxidation predominate, fragmenting the caffeic acid structure and diminishing bioactivity [27]. These findings indicate that drying at ~50 °C is optimal to preserve clovamide, whose antioxidant potency makes it valuable for nutraceutical applications.

Procyanidin B2 peaked at 952 ± 44 mg/kg dm at 40–50 °C, with concentrations significantly lower at 30 °C (670 ± 2 mg/kg dm) and dropping again at 70 °C (670 ± 52 mg/kg dm). The moderate accumulation at 40–50 °C suggests enhanced extractability without critical structural degradation, while losses at 70 °C are attributable to cleavage of the interflavan bond and oxidative conversion into procyanidin A2 and catechin units, as documented in model systems heated at 60–90 °C [37]. These oligomeric flavan-3-ols, although relatively stable compared to clovamide, are prone to oxidative depolymerization under heat, which not only liberates monomeric catechin and epicatechin but also generates insoluble, polymerized residues, reducing the pool of recoverable bioactives.

Epicatechin showed maximal retention at 50 °C (454 ± 21 mg/kg dm), maintaining high levels at 40 °C (340 ± 6 mg/kg dm) and 60 °C (461 ± 22 mg/kg dm) but declining sharply to 319 ± 47 mg/kg dm at 70 °C. Its cis-stereochemistry renders epicatechin particularly prone to epimerization into catechin and to oxidative condensation at elevated temperatures [38]. This epimerization process, accelerated by microwaves or thermal stress, has been reported to yield dimers and trimers, particularly at 60–90 °C in aqueous systems [37]. Such transformations reduce free epicatechin but can transiently enrich monomeric catechin pools.

Catechin exhibited relatively low concentrations across all drying treatments, reaching a maximum of 51 ± 2 mg/kg dm at 40 °C before declining sharply to near-undetectable levels (4 ± 2 mg/kg dm) at 70 °C. Moderate thermal exposure favors catechin formation through the depolymerization of higher-order procyanidins, such as dimers and trimers, under microwave or hot-air conditions [32]. Nevertheless, at temperatures of 70 °C and above, this generation is offset by accelerated degradation through multiple pathways. Catechin can epimerize back to epicatechin or condense into oligomeric species, including dimers, trimers, and dehydrodi(epi)catechin A [37]. Additionally, its monomeric form is highly prone to oxidative polymerization under oxygen-rich hot-air drying, resulting in the formation of insoluble, high-molecular-weight complexes that are not quantified as free catechin [39]. These degradation pathways collectively surpass catechin formation from procyanidin cleavage at high temperatures, explaining the pronounced reduction in its measurable concentration at 70 °C.

Collectively, these results indicate that 40–50 °C represents the optimal drying range for CPH, balancing the extractability of key phenolics with minimal degradation. Higher temperatures (≥60 °C) accelerate non-enzymatic oxidation and interflavan cleavage, while lower temperatures (≤30 °C) favor enzymatic browning, both of which diminish the availability of bioactive compounds.

3. Materials and Methods

3.1. Chemicals and Reagents

Ethanol (analytical grade) for extraction was purchased from Kimia (Witham, UK). LC-MS grade water was obtained from Merck KGaA (Darmstadt, Germany). HPLC-grade acetonitrile was sourced from Fisher Chemical (Mumbai, India). Analytical standards, including catechin, epicatechin, procyanidin B2, and caffeic acid, were supplied by Sigma-Aldrich (St. Louis, MO, USA) and LGC Standards (Toronto, ON, Canada).

3.2. Raw Material

Samples of the CCN-51 cocoa variety were purchased in March 2022 from a local producer in the canton of Ventanas, Los Ríos Province, Ecuador. The pods were thoroughly washed with water to remove surface impurities and allowed to rest for four days to facilitate the natural detachment of the mucilage-coated beans from the pod husk. Each pod was first split longitudinally through the center to remove the beans, and each half was then cut into approximately four sections, resulting in a total of about eight segments per pod. The husk pieces were immediately subjected to the drying process.

3.3. Dehydration Process

CPH was dried at 30 °C, 40 °C, 50 °C, 60 °C, and 70 °C using a hot-air dryer (DY-330H, Lassele, Ansan, Republic of Korea). The samples were periodically turned to ensure uniform exposure to airflow, and drying was continued until the moisture content was reduced to below 10%. The drying times required to reach this target were approximately 62, 50, 39, 30, and 23 h, respectively. The final moisture content of each sample was determined using a moisture analyzer (HE73-03, Mettler Toledo, Columbus, OH, USA). The dried husks were subsequently ground using an ultra-centrifugal mill (ZM 200, Retsch GmbH, Haan, Germany) and sieved with a vibratory sieve shaker (AS 200, Retsch GmbH, Haan, Germany) to obtain particles within a size range of 250–500 μm. All drying experiments were conducted in triplicate. The resulting CPH powder was stored at room temperature, protected from light, until further extraction and analysis.

3.4. Experimental Design

MAE was optimized using RSM. A BBD with three factors and three levels (3^3) was employed, where extraction temperature (50–150 °C), ethanol concentration (0–100%), and extraction time (15–60 min) were selected as the independent variables. For all experimental runs, CPH previously dried at 50 °C was used, and the solid-to-solvent ratio (1:10) and particle size (250–500 μm) were kept constant for all experiments. The concentrations of catechin, epicatechin, procyanidin B2, and clovamide were used as response variables. The experimental design comprised 15 randomized runs, including replicated central points to estimate pure error and evaluate lack of fit, as detailed in Table 2. Randomization was applied to minimize uncontrolled variability. Analysis of variance (ANOVA) followed by Fisher's test was used to determine the statistical significance of main effects and interactions. The adequacy of the fitted models was assessed based on the coefficient of determination (R^2) and lack-of-fit tests.

Table 2. Experimental matrix of the Box–Behnken Design showing the effect of ethanol concentration, extraction temperature, and extraction time on the concentration of phenolic compounds (catechin, epicatechin, procyanidin B2, and clovamide) in cocoa pod husk expressed as mg/kg dry matter obtained by microwave-assisted extraction.

Run	Ethanol (%)	Temperature (°C)	Time (min)	Clovamide (mg/kg dm)	Procyanidin B2 (mg/kg dm)	Epicatechin (mg/kg dm)	Catechin (mg/kg dm)
1	50	100	37.5	3471	929	413	31
2	100	100	60	613	130	199	<LOQ
3	0	100	60	1328	25	101	26
4	100	100	15	888	200	107	<LOQ
5	50	100	37.5	3397	845	466	40
6	0	50	37.5	816	36	12	<LOQ
7	0	100	15	1361	283	96	<LOQ
8	50	100	37.5	3570	988	454	27
9	100	50	37.5	305	54	54	<LOQ
10	0	150	37.5	808	<LOQ	<LOQ	<LOQ
11	50	50	15	2441	518	153	<LOQ
12	100	150	37.5	845	80	308	15
13	50	150	15	238	<LOQ	74	57
14	50	150	60	783	<LOQ	46	44
15	50	50	60	2672	565	160	<LOQ

<LOQ: concentrations below the limit of quantification.

3.5. Microwave-Assisted Extraction

For each extraction, 2 g of previously dried CPH was placed in 20 mL of solvent inside 40 mL Teflon vessels. The 50 °C drying temperature was selected as it effectively reduces moisture content while limiting thermal degradation of phenolic compounds, consistent with previous studies on cocoa by-products [6]. Extractions were carried out in a laboratory microwave system (Mars 6, CEM Corporation, Matthews, NC, USA) operating at 1200 W, with a 10 min ramp period to reach the target extraction temperature. Once the extraction time had elapsed, the system was programmed for a 15 min cooling phase before the vessels were removed. Immediately afterward, the vessels were immersed in an ice-water bath to prevent degradation of thermolabile compounds. The extracts were then transferred to 50 mL centrifuge tubes and centrifuged at 8500 rpm for 20 min using a centrifuge (Sorvall ST8, Thermo Fisher Scientific, Waltham, MA, USA). The resulting supernatants were collected and stored at −20 °C until further analysis.

3.6. Quantification of Phenolic Compounds by HPLC-DAD-ESI-IT-MS/MS

All extracts obtained from the microwave-assisted extraction experiments, as well as samples from each drying temperature evaluated under the optimal extraction conditions, were analyzed by HPLC–DAD–ESI–IT–MS/MS. Prior to chromatographic analysis, supernatants were filtered through 0.2 μm Choice™ regenerated cellulose syringe filters (CH2225-RC, Thermo Scientific, Waltham, MA, USA) to remove particulate matter. Bioactive compounds from the cocoa by-product were analyzed using a high-performance liquid chromatography system (Dionex Ultimate 3000, Thermo Scientific, Waltham, MA, USA) coupled to an ion trap mass spectrometer with an electrospray ionization (ESI) source (AmaZon Speed, Bruker Daltonics, Billerica, MA, USA). Chromatographic separation was performed on a C18 reversed-phase column (Acclaim™ 120, 250 mm × 4.6 mm, 5 μm; Thermo Fisher Scientific, Waltham, MA, USA) maintained at 25 °C, following the procedure described by Maldonado and Figueroa [19]. The mobile phase consisted of acetonitrile (MeCN) and water (H_2O), using a decreasing polarity gradient at a constant flow rate of 0.5 mL/min: 0–3 min, 5% MeCN; 3–10 min, linear increase to 18% MeCN; held at 18% until 19 min; 19–20 min, increased to 21% MeCN; 20–50 min, linear increase to 30% MeCN; 50–60 min, linear increase to 60% MeCN; 60–61 min, held at 60% MeCN; and 61–66 min, decreased to 5% MeCN. The injection volume was 10 μL, and detection was performed with a photodiode array detector (Thermo Scientific, Waltham, MA, USA) at 280 nm.

The system was controlled using Bruker Daltonics (Bremen, Germany) HyStar 3.2 software with an Ethernet data interface. Mass spectrometry was performed in negative ionization mode under the following conditions: capillary voltage 4.500 V; m/z range 100–2000; nebulizer pressure 26.0 psi; nitrogen drying gas temperature 200 °C; drying gas flow 6.0 L/min; rolling averages set to 2 counts; number of averages set to 2; and ion charge control activated.

Catechin, epicatechin, and procyanidin B2 were quantified using external standards, while clovamide was expressed as caffeic acid equivalents due to their structural and ionization similarities, in accordance with established phytochemical practices [40]. Calibration curves for all compounds were constructed with six concentration levels and yielded $R^2 \geq 0.995$. The limit of quantification, determined from standard solutions, was 1.0 mg/kg and corresponded to the lowest concentration that yielded an acceptable coefficient of determination (R^2) in the calibration curve.

3.7. Statistical Analysis

Once the optimal extraction conditions were established, the effect of drying temperature on polyphenol content was evaluated. For this purpose, ANOVA was performed, followed by Tukey's multiple range test for post hoc comparisons. All statistical analyses were conducted using Minitab software, version 14.0 (Minitab Inc., State College, PA, USA).

4. Conclusions

MAE of phenolic compounds from CPH was successfully optimized using response surface methodology, identifying 51% ethanol, 104 °C, and 38 min as the optimal conditions. Under these parameters, clovamide, procyanidin B2, and epicatechin were efficiently recovered, whereas catechin levels were largely influenced by the thermal depolymerization of procyanidins. Drying temperature exerted a significant effect on phenolic retention, with 40–50 °C identified as the optimal range to simultaneously achieve enzymatic inactivation and minimize thermal degradation. These findings establish a robust process framework for the valorization of CPH as a sustainable source of antioxidant-rich phenolics with potential applications in functional foods and nutraceuticals.

Future research should focus on enhancing extract stability and expanding its applications in food systems. Microencapsulation strategies, including spray-drying, freeze-drying, and complex coacervation, should be further explored to protect phenolic compounds during processing and storage. In parallel, alternative approaches such as nanoemulsions, edible coatings, and direct incorporation in powdered form could provide versatile options for functional ingredient development. The extract may also be incorporated into beverages, bakery products, dairy, and confectionery, where phenolics can act as natural antioxidants or bioactive enhancers. Evaluating these applications under realistic processing and storage conditions will offer valuable insight into technological feasibility and functional performance. Additionally, bioaccessibility and bioavailability studies, employing simulated gastrointestinal digestion and Caco-2 cell assays, will be essential to confirm physiological relevance. Another promising avenue is the isolation of clovamide, procyanidin B2, catechin, and epicatechin using flash chromatography or preparative HPLC. This would enable targeted investigations into their bioactivity, stability, synergistic interactions, and bioavailability. Finally, integrating MAE into a sequential biorefinery for the co-recovery of pectin could enhance both economic viability and sustainability in CPH valorization.

Author Contributions: Conceptualization, J.G.F.; methodology, J.G.F.; formal analysis, P.G.; investigation, P.G. and C.R.; resources, J.G.F.; data curation, J.G.F.; writing—original draft preparation, P.G. and C.R.; writing—review and editing, J.G.F.; visualization, P.G.; supervision, J.G.F. All authors have read and agreed to the published version of the manuscript.

Funding: This research was funded by Universidad Técnica Particular de Loja, grant number POA_VIN-56 and VI CONV. TFT 2024. The APC was funded by Universidad Técnica Particular de Loja.

Institutional Review Board Statement: Not applicable.

Informed Consent Statement: Not applicable.

Data Availability Statement: All data used in this study are available upon request from the corresponding author.

Acknowledgments: The authors are very grateful to the Universidad Técnica Particular de Loja (UTPL) for supporting this investigation and open-access publication.

Conflicts of Interest: The authors declare no conflicts of interest. The funders had no role in the design of the study; in the collection, analyses, or interpretation of data; in the writing of the manuscript; or in the decision to publish the results.

Abbreviations

The following abbreviations are used in this manuscript:

ANOVA	Analysis of variance
BBD	Box–Behnken Design
CPH	Cocoa pod husk
DAD	Diode-Array Detector
dm	Dry matter
ESI	Electrospray Ionization
HPLC	High-Performance Liquid Chromatography
IT	Ion Trap
LOQ	Limit of quantification
MAE	Microwave-assisted extraction
MS/MS	Tandem mass spectrometry
m/z	Mass-to-charge ratio
R^2	Coefficients of determination
RSM	Response Surface Methodology
RT	Retention time

References

1. Lima, G.V.S.; e Gonçalves, C.G.; Pinto, A.S.O.; da Silva, E.M.; de Souza, J.N.S.; Rogez, H. Impact of post-harvest processing and roasting conditions on the physicochemical properties, phenolic compounds, and antioxidant capacity of cocoa beans from the Brazilian Amazon. *LWT* **2024**, *210*, 116825. [CrossRef]
2. FAO (Food and Agriculture Organisation of the United Nations). FAOSTAT. Available online: https://www.fao.org/faostat/es#data/QCL (accessed on 2 May 2025).
3. Cádiz-Gurrea, M.d.l.L.; Fernández-Ochoa, Á.; Leyva-Jiménez, F.J.; Guerrero-Muñoz, N.; Villegas-Aguilar, M.d.C.; Pimentel-Moral, S.; Ramos-Escudero, F.; Segura-Carretero, A. LC-MS and spectrophotometric approaches for evaluation of bioactive compounds from Peru cocoa by-products for commercial applications. *Molecules* **2020**, *25*, 3177. [CrossRef] [PubMed]
4. Nguyen, V.T.; Tran, T.G.; Tran, N.L. Phytochemical compound yield and antioxidant activity of cocoa pod husk (*Theobroma cacao* L.) as influenced by different dehydration conditions. *Dry Technol.* **2021**, *40*, 2021–2033. [CrossRef]
5. Mashuni; Hamid, F.H.; Muzuni; Kadidae, L.O.; Jahiding, M.; Ahmad, L.O.; Saputra, D. The determination of total phenolic content of cocoa pod husk based on microwave-assisted extraction method. *AIP Conf. Proc.* **2020**, *2243*, 030013. [CrossRef]
6. Ramos-Escudero, F.; Casimiro-Gonzales, S.; Cádiz-Gurrea, M.d.l.L.; Cancino Chávez, K.; Basilio-Atencio, J.; Ordoñez, E.S.; Muñoz, A.M.; Segura-Carretero, A. Optimizing vacuum drying process of polyphenols, flavanols and DPPH radical scavenging assay in pod husk and bean shell cocoa. *Sci. Rep.* **2023**, *13*, 13900. [CrossRef] [PubMed]
7. Abdul Karim, A.; Azlan, A.; Ismail, A.; Hashim, P.; Abd Gani, S.S.; Zainudin, B.H.; Abdullah, N.A. Phenolic composition, antioxidant, anti-wrinkles and tyrosinase inhibitory activities of cocoa pod extract. *BMC Complement Altern. Med.* **2014**, *14*, 381. [CrossRef]
8. Ramos-Escudero, F.; Rojas-García, A.; Cádiz-Gurrea, M.d.l.L.; Segura-Carretero, A. High potential extracts from cocoa byproducts through sonotrode optimal extraction and a comprehensive characterization. *Ultrason Sonochem.* **2024**, *106*, 106887. [CrossRef]
9. Valadez-Carmona, L.; Plazola-Jacinto, C.P.; Hernández-Ortega, M.; Hernández-Navarro, M.D.; Villarreal, F.; Necoechea-Mondragón, H.; Ortiz-Moreno, A.; Ceballos-Reyes, G. Effects of microwaves, hot air and freeze-drying on the phenolic compounds, antioxidant capacity, enzyme activity and microstructure of cacao pod husks (*Theobroma cacao* L.). *Innov. Food Sci. Emerg. Technol.* **2017**, *41*, 378–386. [CrossRef]
10. ElGamal, R.; Song, C.; Rayan, A.M.; Liu, C.; Al-Rejaie, S.; ElMasry, G. Thermal degradation of bioactive compounds during drying process of horticultural and agronomic products: A comprehensive overview. *Agronomy* **2023**, *13*, 1580. [CrossRef]
11. Fan, F.-Y.; Shi, M.; Nie, Y.; Zhao, Y.; Ye, J.-H.; Liang, Y.-R. Differential behaviors of tea catechins under thermal processing: Formation of non-enzymatic oligomers. *Food Chem.* **2016**, *196*, 347–354. [CrossRef]
12. Osorio-Tobón, J.F. Recent advances and comparisons of conventional and alternative extraction techniques of phenolic compounds. *J. Food Biosci. Technol.* **2020**, *57*, 4299–4315. [CrossRef]
13. Herrero, M.; Martín-Álvarez, P.J.; Senorans, F.J.; Cifuentes, A.; Ibáñez, E. Optimization of accelerated solvent extraction of antioxidants from *Spirulina platensis* microalga. *Food Chem.* **2005**, *93*, 417–423. [CrossRef]
14. Shen, L.; Pang, S.; Zhong, M.; Sun, Y.; Qayum, A.; Liu, Y.; Rashid, A.; Xu, B.; Liang, Q.; Ma, H.; et al. A comprehensive review of ultrasonic assisted extraction (UAE) for bioactive components: Principles, advantages, equipment, and combined technologies. *Ultrason Sonochem.* **2023**, *101*, 106646. [CrossRef]
15. Routray, W.; Orsat, V. Microwave-assisted extraction of flavonoids: A review. *Food Bioproc. Technol.* **2012**, *5*, 409–424. [CrossRef]
16. de Castro, M.L.; Castillo-Peinado, L. Microwave-assisted extraction of food components. In *Innovative Food Processing Technologies Extraction, Separation, Component Modification and Process Intensification*; Elsevier: Duxford, UK, 2016; pp. 57–110.
17. Radoiu, M.; Kaur, H.; Bakowska-Barczak, A.; Splinter, S. Microwave-assisted industrial scale cannabis extraction. *Technologies* **2020**, *8*, 45. [CrossRef]
18. Rebollo-Hernanz, M.; Cañas, S.; Taladrid, D.; Segovia, Á.; Bartolomé, B.; Aguilera, Y.; Martín-Cabrejas, M.A. Extraction of phenolic compounds from cocoa shell: Modeling using response surface methodology and artificial neural networks. *Sep. Purif. Technol.* **2021**, *270*, 118779. [CrossRef]
19. Maldonado, Y.E.; Figueroa, J.G. Microwave-assisted extraction optimization and effect of drying temperature on catechins, procyanidins and theobromine in cocoa beans. *Molecules* **2023**, *28*, 3755. [CrossRef] [PubMed]
20. Rahayu, P.; Rosyidi, D.; Purwadi; Imam, T. Characteristics of catechin extracted from cocoa husks using microwave assisted extraction (MAE). *Biodiversitas* **2019**, *20*, 3626–3631.
21. Van, T.N.; Thi, D.P.; Long, B.V.; Van, H.N.; Ngoc, L.T. Microwave-assisted extraction for maximizing the yield of phenolic compounds and antioxidant capacity from cacao pod husk (*Theobroma cacao* L.). *Curr. Nutr. Food Sci.* **2021**, *17*, 225–237. [CrossRef]
22. Figueroa, J.G.; Borrás-Linares, I.; Lozano-Sánchez, J.; Segura-Carretero, A. Comprehensive characterization of phenolic and other polar compounds in the seed and seed coat of avocado by HPLC-DAD-ESI-QTOF-MS. *Food Res. Int.* **2018**, *105*, 752–763. [CrossRef]
23. Setford, P.C.; Jeffery, D.W.; Grbin, P.R.; Muhlack, R.A. Factors affecting extraction and evolution of phenolic compounds during red wine maceration and the role of process modelling. *Trends Food Sci. Technol.* **2017**, *69*, 106–117. [CrossRef]

24. Moreira, M.M.; Barroso, M.F.; Boeykens, A.; Withouck, H.; Morais, S.; Delerue-Matos, C. Valorization of apple tree wood residues by polyphenols extraction: Comparison between conventional and microwave-assisted extraction. *J. Industrial Crops* **2017**, *104*, 210–220. [CrossRef]
25. Dasiman, R.; Nor, N.M.; Eshak, Z.; Mutalip, S.S.M.; Suwandi, N.R.; Bidin, H. A review of procyanidin: Updates on current bioactivities and potential health benefits. *Biointerface Res. Appl. Chem.* **2022**, *12*, 5918–5940.
26. Isidoro Haminiuk, C.W.; Vicente Plata-Oviedo, M.S.; de Mattos, G.; Carpes, S.T.; Branco, I.G. Fundamental and applied aspects of catechins from different sources: A review. *Int. J. Food Sci. Technol.* **2020**, *55*, 429–442.
27. Locatelli, M.; Travaglia, F.; Giovannelli, L.; Coïsson, J.D.; Bordiga, M.; Pattarino, F.; Arlorio, M. Clovamide and phenolics from cocoa beans (*Theobroma cacao* L.) inhibit lipid peroxidation in liposomal systems. *Food Res. Int.* **2013**, *50*, 129–134. [CrossRef]
28. Tranquilino-Rodríguez, E.; Martínez-Flores, H.E.; Rodiles-López, J.O.; Dios Figueroa-Cárdenas, J.D.; Pérez-Sánchez, R.E. Optimization in the extraction of polyphenolic compounds and antioxidant activity from *Opuntia ficus-indica* using response surface methodology. *J. Food Process Preserv.* **2020**, *44*, e14485. [CrossRef]
29. Zhang, Z.; Lv, G.; Pan, H.; Fan, L. Optimisation of the microwave-assisted extraction process for six phenolic compounds in *Agaricus blazei* Murrill. *Int. J. Food Sci. Technol.* **2012**, *47*, 24–31. [CrossRef]
30. Vo, T.P.; Pham, T.V.; Tran, T.N.H.; Vo, L.T.V.; Vu, T.T.; Pham, N.D.; Nguyen, D.Q. Ultrasonic-assisted and microwave-assisted extraction of phenolics and terpenoids from *Abelmoschus sagittifolius* (Kurz) Merr roots using natural deep eutectic solvents. *ACS Omega* **2023**, *8*, 29704–29716. [CrossRef]
31. Liazid, A.; Palma, M.; Brigui, J.; Barroso, C.G. Investigation on phenolic compounds stability during microwave-assisted extraction. *J. Chromatogr. A* **2007**, *1140*, 29–34. [CrossRef]
32. Chen, J.; Thilakarathna, W.W.; Astatkie, T.; Rupasinghe, H.V. Optimization of catechin and proanthocyanidin recovery from grape seeds using microwave-assisted extraction. *Biomolecules* **2020**, *10*, 243. [CrossRef] [PubMed]
33. Multari, S.; Marsol-Vall, A.; Keskitalo, M.; Yang, B.; Suomela, J.-P. Effects of different drying temperatures on the content of phenolic compounds and carotenoids in quinoa seeds (*Chenopodium quinoa*) from Finland. *J. Food Compost. Anal.* **2018**, *72*, 75–82. [CrossRef]
34. Yener, E.; Saroglu, O.; Sagdic, O.; Karadag, A. The effects of different drying methods on the *in vitro* bioaccessibility of phenolics, antioxidant capacity, and morphology of European plums (*Prunes domestica* L.). *ACS Omega* **2024**, *9*, 12711–12724. [CrossRef]
35. Oracz, J.; Nebesny, E.; Żyżelewicz, D. Changes in the flavan-3-ols, anthocyanins, and flavanols composition of cocoa beans of different *Theobroma cacao* L. groups affected by roasting conditions. *Eur. Food Res. Technol.* **2015**, *241*, 663–681. [CrossRef]
36. Mesquita, V.L.V.; Queiroz, C. Enzymatic browning. In *Biochemistry of Foods*; Eskin, N.A.M., Shahidi, F., Eds.; Academic Press: London, UK, 2013; Volume 3, pp. 387–418.
37. De Taeye, C.; Kankolongo Cibaka, M.-L.; Jerkovic, V.; Collin, S. Degradation of (−)-epicatechin and procyanidin B2 in aqueous and lipidic model systems. First evidence of "chemical" flavan-3-ol oligomers in processed cocoa. *J. Agric. Food Chem.* **2014**, *62*, 9002–9016. [CrossRef]
38. Wang, R.; Zhou, W.; Jiang, X. Reaction kinetics of degradation and epimerization of epigallocatechin gallate (EGCG) in aqueous system over a wide temperature range. *J. Agric. Food Chem.* **2008**, *56*, 2694–2701. [CrossRef] [PubMed]
39. Latos-Brozio, M.; Masek, A. Natural polymeric compound based on high thermal stability catechin from green tea. *Biomolecules* **2020**, *10*, 1191. [CrossRef] [PubMed]
40. Mena, P.; Cirlini, M.; Tassotti, M.; Herrlinger, K.A.; Dall'Asta, C.; Del Rio, D. Phytochemical profiling of flavonoids, phenolic acids, terpenoids, and volatile fraction of a rosemary (*Rosmarinus officinalis* L.) extract. *Molecules* **2016**, *21*, 1576. [CrossRef] [PubMed]

molecules

MDPI

Article

Exploring the Influence of a Pomegranate Extract on the Functionality of Healthy and Diseased Human Gut Microbiota: An In Vitro Study

Daniele Giuseppe Buccato [1], Adriana Delgado-Osorio [2], Lorenza Francesca De Lellis [1], Maria Vittoria Morone [3], Hammad Ullah [4], Luana Izzo [1], Sonia Lombardi [1], Alessandro Di Minno [1,5], Costanza Valentina Riccioni [6], Dafni Moriki [7], José Ángel Rufián-Henares [2,8] and Maria Daglia [1,9,*]

1 Department of Pharmacy, University of Napoli Federico II, Via D. Montesano 49, 80131 Naples, Italy; danielegiuseppe.buccato@unina.it (D.G.B.); lorenzafrancesca.delellis@unina.it (L.F.D.L.); luana.izzo@unina.it (L.I.); sonia.lombardi@unina.it (S.L.); alessandro.diminno@unina.it (A.D.M.)
2 Department of Nutrition and Food Sciences, Institute of Nutrition and Food Technology, Biomedical Re-Search Centre, Universidad de Granada, 18071 Granada, Spain; adrianadelgado@ugr.es (A.D.-O.); jarufian@ugr.es (J.Á.R.-H.)
3 Department of Experimental Medicine, Section of Microbiology and Clinical Microbiology, University of Campania "L. Vanvitelli", 80138 Naples, Italy; mariavittoria.morone@unicampania.it
4 School of Pharmacy, University of Management and Technology, Lahore 54000, Pakistan; hammadrph@gmail.com
5 CEINGE-Biotecnologie Avanzate, Via Gaetano Salvatore 486, 80145 Naples, Italy
6 R&D Department, Esserre Pharma Srl, 00191 Rome, Italy; c.riccioni@esserrepharma.it
7 Allergology and Pulmonology Unit, 3rd Pediatric Department, National and Kapodistrian University of Athens, 12462 Athens, Greece; dafnimoriki@yahoo.gr
8 Instituto de Investigación Biosanitaria (ibs.GRANADA), Universidad de Granada, Granada 18140, Spain
9 International Research Center for Food Nutrition and Safety, Jiangsu University, Zhenjiang 212013, China
* Correspondence: maria.daglia@unina.it

Abstract: Pomegranate is recognized for its health benefits, primarily due to its polyphenols and metabolites, such as urolithins (Uro-A), produced via colonic fermentation of ellagic acid (EA). These compounds make pomegranate a functional food with the potential to modulate chronic disease risk factors and enhance gut health by modulating microbiota. The aims of this study were (1) to evaluate the effect of in vitro digestion and fermentation mimicking human digestive processes on the total phenolic content (TPC) and the antioxidant capacity of a standardized pomegranate extract (PE), (2) to assess the effect of the digested PE on the functionality of gut microbiota isolated from healthy and diseased subject fecal materials via short-chain fatty acid (SCFA) determination, and (3) to measure Uro-A production using UHPLC Q-Orbitrap HRMS. The in vitro digestion and fermentation processes resulted in a significant increase in the TPC, while the antioxidant capacity was considerably reduced. Following the in vitro digestion process, the TPC increased from 232 ± 16 to 1656 ± 34 g GAE/g of PE. Moreover, the TPC in the fermented samples was calculated as 6139 ± 458 g GAE/g for the microbiota of healthy adults and 8375 ± 1388 g GAE/g for the microbiota of healthy children, compared to 1657 ± 34 g GAE/g for the non-fermented samples. The PE exerted a modulatory effect on gut microbiota functionality, as reflected by an increasing concentration of SCFAs, especially lactic acid. Overall, these data suggest that pomegranate might contribute to gut health and could be a candidate for further studies in view of its possible use as a prebiotic ingredient. Further research, including clinical studies, is needed to confirm these findings and explore the potential application of pomegranate extract as a functional ingredient in nutraceuticals and functional foods aimed at improving gut health.

Keywords: pomegranate extract; polyphenols; in vitro digestion; in vitro fermentation; antioxidant capacity; short-chain fatty acids; urolithins

https://doi.org/10.3390/molecules30071634

1. Introduction

Pomegranate (*Punica granatum* L., Punicaceae) is a fruit consumed worldwide, native to China, the Indian subcontinent, Afghanistan, and Iran, where historic texts claim that it has been grown in these regions for thousands of years. Nowadays, pomegranate cultivation is extended from Iran, over the Mediterranean region, to the Turkish borders of Europe, the American Southwest, California, and Mexico [1,2]. Data from the U.S. Department of Agriculture [3] indicate that pomegranate fruit (per 100 g) contains dietary fiber (4 g), essential minerals, including calcium (10 mg), iron (0.3 mg), magnesium (12 mg), phosphorus (36 mg), potassium (236 mg), sodium (3 mg), zinc (0.35 mg), copper (0.158 mg), manganese (0.119 mg), and selenium (0.5 μg), as well as a range of vitamins, such as ascorbic acid (10.2 mg), thiamine (0.067 mg), riboflavin (0.053 mg), niacin (0.293 mg), pantothenic acid (0.377 mg), pyridoxine (0.075 mg), folate (38 μg), choline (7.5 mg), alpha-tocopherol (0.6 mg), and phylloquinone (16.4 μg). Additionally, it provides monounsaturated fatty acids (0.093 g) and polyunsaturated fatty acids (0.079 g). Moreover, pomegranate fruits are rich in polyphenols, with hydrolysable tannins (i.e., gallotannins and ellagitannins) being the most abundant ones, followed by hydroxycinnamic acids, flavanols (epigallocatechin), flavanones (naringenin and hesperidin), flavonols (myricetin, quercetin, and kaempferol), flavones (apigenin and luteolin), anthocyanins (delphinidin, cyanidine, and pelargonidine), and dihydrochalcones (phloridzin) [4,5]. These pomegranate phytochemicals, which account for 92% of the antioxidant capacity associated with the fruit [6], may influence cell signaling pathways, scavenge free radicals, chelate metals, and modify gene expression, among other properties [7–9]. In biological systems, pomegranate polyphenols can disrupt free radical chain reactions by donating H^+ to lipid peroxides [10]. Moreover, some compounds (i.e., punicalagin) have the ability to scavenge hydrogen peroxide (H_2O_2) by donating an electron from the phenolic group to H_2O_2 and converting it into H_2O [11]. The ability of pomegranate extracts to reduce Fe^{3+} to Fe^{2+} further demonstrates their electron donor capacity, which is mainly dependent on the quantity of phenolic hydroxyl groups in each molecule [12]. A study showed that quercetin and punicalagin in pomegranate juice chelate intracellular iron, hence limiting the production of free radicals [11]. Pomegranate juice demonstrated the highest antioxidant potency as compared to other polyphenol-rich beverages such as orange juice, red wine, green and black tea, cranberry juice, black cherry juice, blueberry juice, and açaí juice [12]. Other interesting biological effects of pomegranate fruit include anti-inflammatory properties against diseases induced by chronic low-grade systemic inflammation (i.e., rheumatoid arthritis, asthma, chronic obstructive pulmonary disease, or inflammatory bowel disease), cardiometabolic disorders (obesity, diabetes mellitus, hypertension, hypercholesterolemia, and atherosclerosis), different types of cancer, microbial infections, and wound healing [6,13].

Ellagitannins, especially punicalagins, obtained from pomegranate byproducts suppress the growth of certain pathogenic bacteria, including *Pseudomonas aeruginosa*, *Clostridium* species, and *Staphylococcus aureus*. However, they can also promote the growth of commensal bacteria and the de novo synthesis of short-chain fatty acids (SCFAs) such as acetate, propionate, and butyrate [14,15]. Following absorption, SCFAs may be transported to different organs and alter numerous pathways associated with health and disease. Acetate and butyrate are mostly involved in lipid biosynthesis, whereas propionate is primarily involved in gluconeogenesis. In addition to their direct role as significant energy sources, SCFAs can regulate multiple biological host responses, mainly oxidative and inflammatory stresses [16,17]. Animal studies indicated activation of peroxisome proliferator-activated receptors (PPARs) by SCFAs, which control gene expression related to differentiation, metabolism, and cancer [18–20]. By activating PPARα, SCFAs may prevent the expression of proinflammatory mediators, including activator protein (AP)-1, signal transducer and activator of transcription (STAT), and nuclear factor kappa B (NF-κB). Inhibition of these molecules may downregulate the genetic transcription of tumor necrosis factor (TNF-α), interleukins (IL-1 and IL-6), and cyclooxygenase (COX) [21].

This study aimed to evaluate the in vitro impact of a whole fruit pomegranate extract, whose metabolic profile was studied in our previous research [4], on the functionality of gut microbiota obtained from the fecal material of different donors (adults and children, both healthy and diseased, including obese and celiac subjects) using simulated in vitro digestion and fermentation processes, designed to mimic digestion in the human oral, gastric, and intestinal tract.

2. Results

The TPC and antioxidant properties of pomegranate extracts were assessed before and after digestion and fermentation. The antioxidant activity was evaluated using ABTS+ and DPPH assays, which measure radical scavenging capacity, along with the FRAP assay, which determines the reducing potential. The combined use of these assays allowed for a more comprehensive characterization of the antioxidant potential of the pomegranate extract [22]. To identify statistically significant variations between different treatment conditions, the TPC, radical scavenging activity (DPPH and ABTS+), and reducing power (FRAP) were compared across samples subjected to digestion and fermentation with gut microbiota isolated from the fecal material of both healthy and diseased (celiac and obese) individuals, including both adults and children. Additionally, the influence of digested and fermented pomegranate extract on gut microbiota functionality was assessed by quantifying SCFAs using UHPLC-RID analysis. Lastly, the potential of gut microbiota from different donors to facilitate the biotransformation of pomegranate polyphenols into urolithins, particularly Uro-A, was analyzed through UHPLC Q-Orbitrap HRMS.

2.1. TPC Evaluation of Pomegranate Extract Before and After Digestion and Fermentation Processes

The Folin–Ciocalteu assay revealed a significant increase in polyphenol release from the pomegranate extract following in vitro simulated digestion compared to the non-digested samples (Table 1). Specifically, the total phenolic content (TPC) rose from 232 ± 16 to 1656 ± 34 g GAE/g after digestion ($p < 0.0001$). Further fermentation of these digested samples with fecal microbiota from both healthy and diseased individuals led to an even greater increase in the TPC compared to the non-fermented samples (Table 2). Notably, the TPC reached 6139 ± 458 g GAE/g in the samples fermented with microbiota from healthy adults ($p < 0.0001$) and 8375 ± 1388 g GAE/g with microbiota from healthy children ($p < 0.001$), whereas the non-fermented samples remained at 1657 ± 34 g GAE/g. These findings suggest that fermentation with gut microbiota, particularly from healthy individuals, significantly enhances the release of phenolic compounds, underscoring the substantial role of microbial metabolism in polyphenol biotransformation.

Table 1. Antioxidant capacity of the pomegranate extract before and after in vitro oro–gastro–intestinal digestion process.

Assay	Before In Vitro Digestion	After In Vitro Digestion	*p*-Value
TPC (g of GAE/kg of extract)	232 ± 16	1657 ± 34	<0.0001
$TEAC_{DPPH}$ (g of Trolox/kg of extract)	1996 ± 54	2504 ± 23	0.0013
$TEAC_{ABTS}$ (g of Trolox/kg of extract)	4994 ± 127	2198 ± 76	0.0001
$TEAC_{FRAP}$ (g of Trolox/kg of extract)	1924 ± 31	2491 ± 175	0.0373

Values are expressed as the means ± standard deviation. The TPC is expressed in mg of GAE/g of extract, while $TEAC_{DPPH}$, $TEAC_{FRAP}$, and $TEAC_{ABTS}$ are expressed in mg Trolox/g of extract.

Table 2. Antioxidant capacity of the pomegranate extract before and after in vitro digestion and fermentation with gut microbiota isolated from healthy, obese, and celiac subjects.

Assay	In Vitro Digested Samples	After In Vitro Fermentation					
		Adults			Children		
		Healthy	Obese	Celiac	Healthy	Obese	Celiac
TPC	1657 ± 34	6139 ± 458 ****	5889 ± 259 ns	2914 ± 367 ****	8375 ± 1388 ***	8945 ± 1120 ns	8195 ± 1180 ns
$TEAC_{DPPH}$	2504 ± 23	175 ± 11 ****	233 ± 30 ns	241 ± 37 ns	198 ± 31 ****	195 ± 31 ns	203 ± 31 ns
$TEAC_{ABTS}$	2198 ± 76	162 ± 29 ****	107 ± 29 ns	155 ± 23 ns	199 ± 21 ****	193 ± 23 ns	200 ± 31 ns
$TEAC_{FRAP}$	2491 ± 175	251 ± 15 ****	201 ± 18 *	188 ± 21 **	281 ± 28 ****	284 ± 31 ns	267 ± 10 ns

Values are expressed as the means ± standard deviation. The TPC is expressed in mg of GAE/g of extract, while $TEAC_{DPPH}$, $TEAC_{FRAP}$, and $TEAC_{ABTS}$ are expressed in mg Trolox/g of extract. The samples fermented using fecal microbiota isolated from healthy adults and healthy children were compared with the non-fermented samples to determine the statistically significant change in the TPC values and antioxidant profile. Similarly, the samples fermented using fecal microbiota isolated from obese and celiac subjects were compared with those from healthy adults and children to evaluate the impact of the pomegranate extract on the TPC values and antioxidant capacity in diseased states. Note: ns, non-significant, * $p < 0.05$, ** $p < 0.01$, *** $p < 0.001$, **** $p < 0.0001$.

To assess the impact of fermentation with gut microbiota from healthy and diseased individuals on the TPC of the pomegranate extract, a comparative analysis was conducted across different treatment groups. No significant differences were observed in the TPC of the pomegranate extract fermented with microbiota from diseased individuals, except for the samples fermented with microbiota from celiac adults, which showed a statistically significant increase ($p < 0.0001$).

2.2. Radical Scavenging Activity of the Pomegranate Extract Before and After Digestion and Fermentation Processes

The radical scavenging capacity of the pomegranate extract before and after digestion and fermentation was evaluated using DPPH and ABTS+ assays. In its non-digested form, the extract exhibited antioxidant activity equivalent to 1996 ± 54 mg Trolox/g ($TEAC_{DPPH}$) and 4994 ± 127 mg Trolox/g ($TEAC_{ABTS}$). As shown in Table 1, in vitro digestion led to a significant increase in $TEAC_{DPPH}$ (2504 ± 23 mg Trolox/g, $p = 0.0013$) and a notable decrease in $TEAC_{ABTS}$ (2198 ± 76 mg Trolox/g, $p = 0.0001$).

Following in vitro fermentation, both $TEAC_{DPPH}$ and $TEAC_{ABTS}$ values significantly declined compared to the non-fermented samples ($p < 0.0001$, Table 2). Specifically, $TEAC_{DPPH}$ dropped from 2504 ± 23 mg Trolox/g to 175 ± 11 mg Trolox/g after fermentation with microbiota from healthy adults and to 198 ± 31 mg Trolox/g with microbiota from healthy children. Similarly, $TEAC_{ABTS}$ decreased from 2198 ± 76 mg Trolox/g to

162 ± 29 mg Trolox/g and 199 ± 21 mg Trolox/g, respectively. Notably, microbiota from diseased individuals (obese and celiac subjects) did not significantly alter $TEAC_{DPPH}$ or $TEAC_{ABTS}$ values compared to those observed after fermentation with microbiota from healthy adults and children.

2.3. Reducing Power of the Pomegranate Extract Before and After Digestion and Fermentation Processes

The FRAP assay indicated that the non-digested pomegranate extract exhibited a $TEAC_{FRAP}$ value of 1924 ± 30.8 mg Trolox/g. As shown in Table 1, simulated in vitro digestion led to a slight but significant increase in $TEAC_{FRAP}$ (2491 ± 175 mg Trolox/g, $p = 0.0373$). However, in vitro fermentation resulted in a significant reduction in $TEAC_{FRAP}$ compared to the non-fermented samples ($p < 0.0001$, Table 2). Specifically, $TEAC_{FRAP}$ decreased to 251 ± 15.3 mg Trolox/g in the samples fermented with microbiota from healthy adults and to 281 ± 27.7 mg Trolox/g in those fermented with microbiota from healthy children.

Among the diseased subjects, fermentation with microbiota from obese (201 ± 17 mg Trolox/g, $p < 0.05$) and celiac adults (188 ± 21 mg Trolox/g, $p < 0.01$) resulted in significantly lower $TEAC_{FRAP}$ values compared to the samples fermented with microbiota from healthy adults (251 ± 15 mg Trolox/g). In contrast, no significant differences were observed in $TEAC_{FRAP}$ between the samples fermented with microbiota from obese children (284 ± 31.3 mg Trolox/g) or celiac children (267 ± 10 mg Trolox/g) compared to their healthy counterparts (281 ± 27.7 mg Trolox/g).

2.4. SCFA Quantification After In Vitro Digestion and Fermentation

The analysis of SCFA production by gut microbiota isolated from healthy, obese, and celiac subjects in response to the digested pomegranate extract revealed significant differences between adults and children, likely reflecting variations in gut microbiota composition between these age groups. Notably, the gut microbiota of celiac and obese adults produced significantly higher levels of lactic acid following the addition of the digested pomegranate extract compared to healthy adults, whose lactate levels remained relatively unchanged (Figure 1). This suggests that pomegranate extract does not substantially affect lactate production in the gut microbiota of healthy adults compared to the control sample (i.e., fecal material without the pomegranate extract).

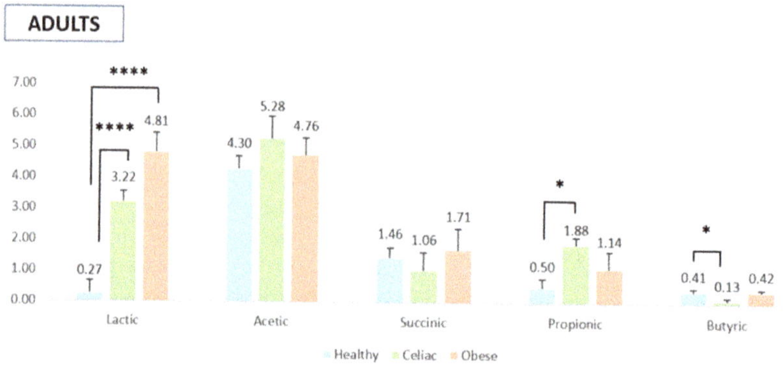

Figure 1. Production of SCFAs by the gut microbiota of adults upon addition of the digested pomegranate extract. The values are expressed in mmol/L as the net increase in the SCFA levels, determined by subtracting the SCFA values of the blank controls (fecal samples fermented without the pomegranate extract) from those of the pomegranate-containing samples. Data are presented as the means ± SD. Note: * $p < 0.05$ and **** $p < 0.0001$.

Although no statistically significant differences were observed among the three adult groups (healthy, obese, and celiac), a slight increase in acetic acid production and, to a lesser extent, succinic acid was noted after pomegranate extract supplementation. Additionally, the extract enhanced propionic acid production in celiac and obese subjects, with a statistically significant increase observed only between celiac and healthy adults. Lastly, a minor elevation in butyric acid levels was recorded in the microbiota of healthy and obese adults compared to the control samples (Figure 1).

Figure 2 presents the SCFA production patterns in the pediatric population. The addition of the digested pomegranate extract during fermentation resulted in a significant increase in lactic acid production, whereas the levels of other SCFAs remained largely unchanged. Notably, acetic acid concentrations were higher in the samples treated with gut microbiota from healthy children compared to those from the children with pathological conditions. Regarding succinic and propionic acids, the gut microbiota of obese children exhibited significantly higher production levels compared to the reference group. However, statistical analysis did not reveal significant differences in butyric acid production among the three experimental groups.

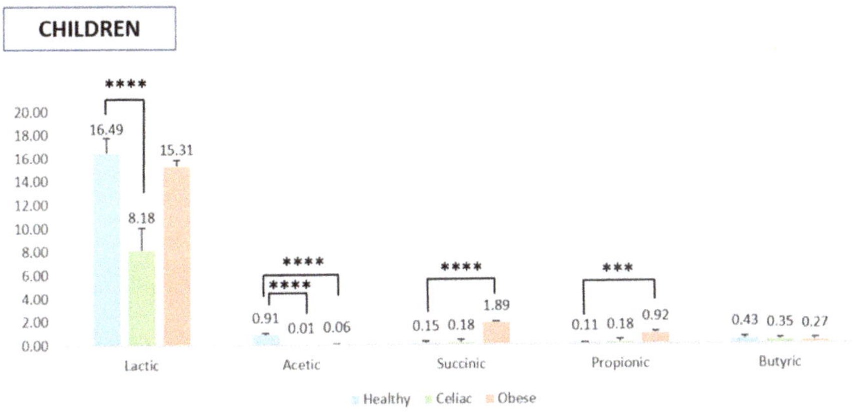

Figure 2. Production of SCFAs by the gut microbiota of children upon addition of the digested pomegranate extract. Values are expressed in mmol/L as the net increase in the SCFA levels, determined by subtracting the SCFA values of the blank controls (fecal samples fermented without the pomegranate extract) from those of the pomegranate-containing samples. Data are presented as the means ± SD. Note: *** $p < 0.001$ and **** $p < 0.0001$.

2.5. Quali–Quantitative Analysis of Urolithin A After In Vitro Fermentation

The quantification of Uro-A was performed using a calibration curve prepared with a pure standard (purity $\geq$ 90%) dissolved in methanol containing 0.1% formic acid, with a concentration range of 0.01 to 5.0 mg/L. Data are presented in Figures 3 and 4.

The UHPLC Q-Orbitrap HRMS analysis was conducted on a total of 36 samples to identify and quantify the presence of urolithins in the fermentation liquids produced by the gut microbiota of healthy, obese, and celiac subjects. The results revealed that only Uro-A was detected in one group of samples (digested pomegranate extract treated with gut microbiota isolated from healthy adults) at a concentration of 4.2 ng/mg.

The detection of Uro-A only in the samples from healthy subjects implies that in these experimental conditions, the ability to produce this metabolite may be limited or absent in obese and celiac subjects and in children.

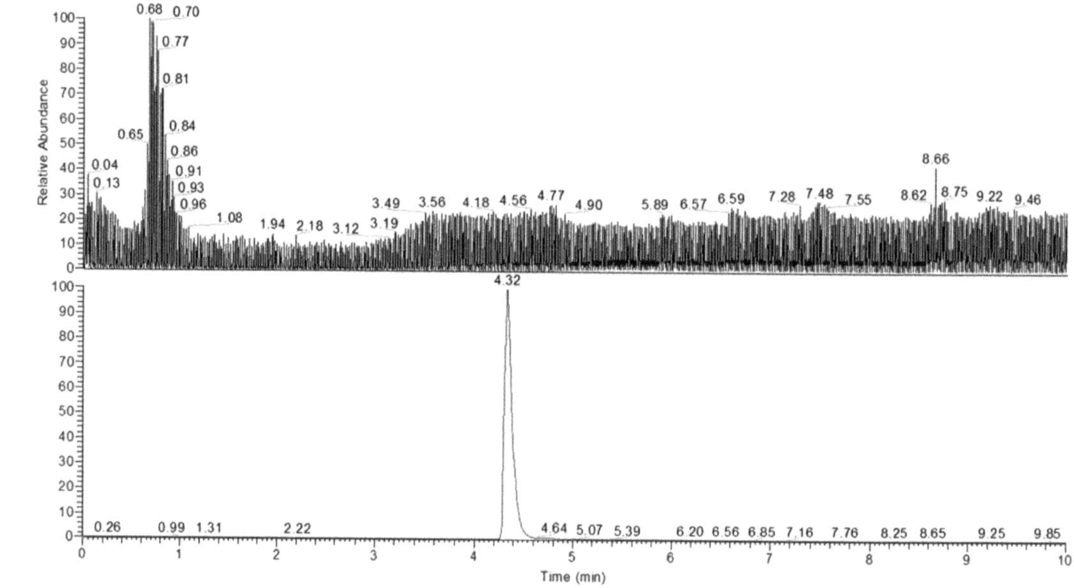

Figure 3. Total ion chromatogram (TIC) of the Uro-A standard obtained using UHPLC-Q-Orbitrap HRMS and extracted ion chromatogram (EIC) of Uro-A, identified at 4.32 min, corresponding to its exact mass (m/z = 227.03465).

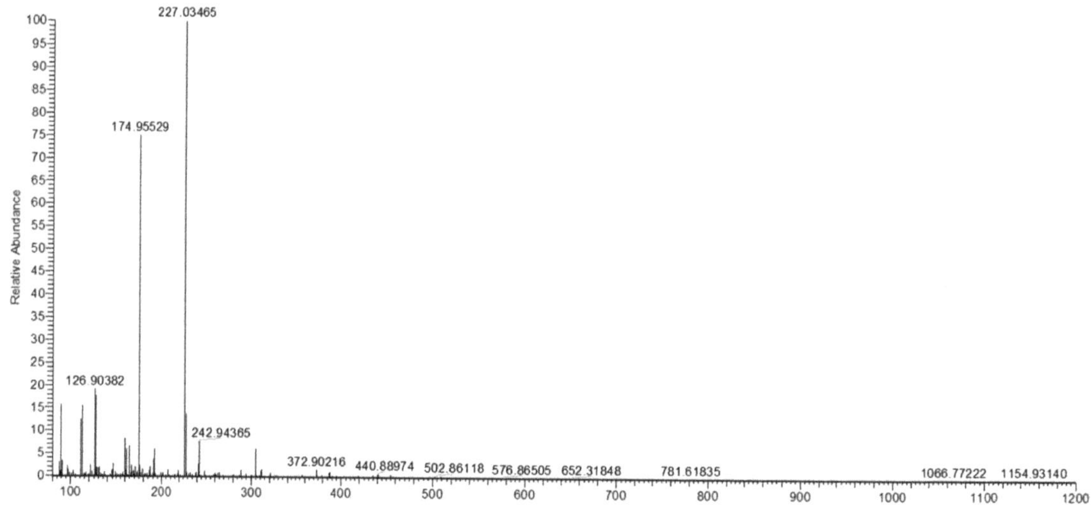

Figure 4. Full mass spectrum acquired in the scan range of m/z 80–500 in the negative ion mode (ESI–) of the Uro-A standard corresponding to m/z 227.03465.

3. Discussion

In this study, a pomegranate extract was subjected to simulated in vitro digestion using predefined protocols, followed by an assessment of its antioxidant profile (i.e., TPC, radical scavenging activity, and reducing power). The digested sample was then fermented with gut microbiota isolated from the fecal material of six donor groups (healthy adults and children, obese adults and children, and celiac adults and children). The digested and fermented samples were analyzed for their antioxidant profile and their capacity to modulate gut microbiota functionality in terms of SCFA production. Additionally, Uro-A quantification was performed after the in vitro fermentation process. The results of these experiments were compared with those obtained from the control samples prepared using the same digestion and fermentation protocols without the addition of the pomegranate extract.

As far as the TPC is concerned, our results are aligned with those obtained by Derakhshan et al. [23], who studied the TPC in ethanolic extracts from different parts of pomegranate, and Ardekani et al. [24], who analyzed the TPC in nine different pomegranate cultivars. The in vitro digestion of the pomegranate extract led to a significant increase in the TPC. These results are in accordance with those obtained by Fawole et al. [25] who observed a significant increase in the TPC following in vitro digestion of pomegranate juice, especially during the gastric phase. These results agree with previous studies on other fruits, showing the same pattern of increase in the TPC following in vitro digestion [26–28]. It can be assumed that the increase in the TPC after in vitro digestion could be attributed to the acidic hydrolysis of phenolic glycosides to their aglycones [29], as previous research on pomegranate extracts showed that most of the polyphenols in the extracts are in glycosidic forms [4]. In addition, pomegranate extract is rich in hydrolysable tannins, which are known to bind tightly with dietary fibers [30]. Digestion processes help to release phenolic compounds from the food matrix (such as fibers) and make them accessible for absorption from the gastrointestinal tract or to be metabolized by gut microbiota [31,32]. Contrarily, Perez-Vicente et al. [33] did not observe any significant effect of the gastric and duodenal digestion process on the TPC in the pomegranate extract. This result might be related to the difference in food matrix characteristics (i.e., pomegranate juice) and experimental conditions for in vitro digestion. Mosele et al. [34] observed an increase in the TPC for the pomegranate extract and a slight decrease in the TPC for the pomegranate juice, suggesting the impact of the food matrix on polyphenol bioaccessibility.

Regarding the antioxidant profile, simulated digestion showed a slight increase in the $TEAC_{DPPH}$ and $TEAC_{FRAP}$ values, while the $TEAC_{ABTS}$ values were significantly decreased. In the present study, it was found that the antioxidant capacity of pomegranate was not correlated with the TPC of the extract. In the literature, it is well-established that a plant extract's antioxidant activity is not only dependent on the polyphenol content, but also on their structure (i.e., number of hydroxyl groups, number and nature of substitutions on rings B and C, and degree of polymerization) [35,36].

With regard to the effect of fermentation on the bioaccessibility of pomegranate polyphenols, to the best of our knowledge, no data are available on the fermentation of pomegranate extract by human gut microbiota obtained from fecal material of healthy and diseased subjects. The present study showed a significant increase in the TPC of the digested pomegranate extract after fermentation with the fecal samples of healthy and diseased subjects. These results were expected, as during the fermentation process, key fermentable microbes can release enzymes such as tannases, glycosidases, esterases, and phenolic acid decarboxylases, which support the release of bound phenolic compounds from the food matrix, thus increasing the concentration of free polyphenol compounds, which are detectable with the Folin–Ciocalteu assay [37]. In such fruits as pomegranate, gut microbes could degrade ellagitannins to some intermediates, such as punicalin and gallagic acid, and eventually into EA [38]. Some fermentative strains (such as *Streptococcus thermophilus*) are known to transform EA to a new metabolite Uro-A, that may increase the TPC [39]. Unlike the TPC, the results of the antioxidant capacity of the pomegranate extract after fermentation with the fecal samples from healthy and diseased subjects were not as expected. A significant decrease in $TEAC_{DPPH}$, $TEAC_{ABTS}$, and $TEAC_{FRAP}$ was observed despite a significant increase in the TPC. Singh et al. [40] reported that the interaction between epithelial cells and commensal bacteria increases the rapid production of ROS to maintain gut barrier protection. Therefore, fecal materials obtained from healthy and diseased adults and children may contain high levels of ROS, which could be scavenged by the anti-radical compounds present in digested pomegranate extract. After interacting with the fecal material, the extract showed a lower residual anti-radical activity than that observed post-digestion alone, likely due to the reduction of ROS levels.

As far as SCFAs are concerned, this study demonstrated that fermentation of the in vitro digested pomegranate extract modulated SCFA production by gut microbiota isolated from diseased subjects, particularly lactic acid in both adults and children. Fermentation with microbiota isolated from adult subjects significantly increased the concentration of lactic acid in diseased individuals (i.e., celiac and especially in obese adults) as compared to healthy ones. In addition, slight effects were observed on acetic and succinic acids when microbiota from healthy, obese, and celiac subjects was treated with the digested pomegranate extract. Notably, fermentation showed statistically significant effects on propionic acid in celiac subjects compared to healthy adults, while a slight increase in butyric acid concentration was noted in healthy and obese subjects. In the microbiota of the obese children, fermentation of the pomegranate extract resulted in a significant increase in lactic, succinic, and propionic acid production. The acetic concentration was increased only in the healthy subjects, while no significant effects on butyric acid were observed in both groups. It is well-known that the gut microbiota plays a crucial role in the fermentation of some dietary compounds, which are inert to digestive enzymes in the upper gastrointestinal tract and are converted into SCFAs (acetate, propionate, and butyrate). In fact, some of these SCFAs are used by enterocytes as a source of energy and exert beneficial effects (i.e., regulatory effects on immunity, metabolism, and gastrointestinal physiology) [41]. Specifically, butyrate promotes intestinal cell regeneration and protection, mucus production, a decrease in hypercholesterolemia levels, and the release of hormones and/or neurotransmitters that control intestinal motility and insulin resistance [42]. In addition to dietary fibers, polyphenols have been recognized as a new class of prebiotics that meet the criteria to be classified as a prebiotic substrate, such as resistance to host digestion, the capacity to be fermented by the gut microbiota, and the ability to modulate gut microbiota functionality [43]. The effect of pomegranate extract on the modulation of the synthesis of SCFAs can be attributed to its high content of polyphenols.

Moreover, the interaction of the gut microbiota with the immune system of the host is one of the key elements explaining the pathogenesis of a large spectrum of diseases, including obesity and celiac disease [44]. Obesity is generally associated with reduced microbial diversity, SCFA-synthesizing bacteria, and SCFA concentration, thus influencing adipose tissue composition, fat mass gain, chronic low-grade inflammation, and insulin resistance [45]. Some in vivo studies suggest that supplementation with SCFAs may reduce body weight, improve insulin sensitivity, and reduce chronic low-grade inflammation [46–50]. Similarly, patients with celiac disease exhibit specific gut microbiota profiles, characterized by reduced microbial diversity and alterations in SCFA concentrations, which may reflect impaired fermentation processes and contribute to intestinal inflammation [51–53]. A decrease in *Bacteroides, Bifidobacterium* species, and *Parabacteroides* in celiac disease, with an increased abundance of *Escherichia, Helicobacter*, and *Candida albicans*, has been reported in the literature [54]. Considering this evidence, gut microbiota modulation can be considered one of the most important targets in the prevention and treatment of chronic pathologies. The current study, with notable effects on lactic acid production, indicated that pomegranate extract may have pronounced effects on the activity of lactic acid-producing bacteria [55]. A study by Mosele et al. [56] showed no significant effects of pomegranate juice on the gut microbiome composition in healthy adults, though the authors observed effects on health-promoting phenolic metabolites, such as catechol and 3-phenylpropionic acid. Another study demonstrated that the fermentation of fecal samples from healthy subjects with pomegranate byproducts resulted in an enhanced growth of *Lactobacillus* and *Bifidobacterium* species, as well as an increased production of SCFAs, such as acetate, propionate, and butyrate, indicating potential prebiotic benefits of these byproducts [14]. In an animal study, Song et al. demonstrated attenuation of obesity, improvement of insulin resistance, and modulation of gut microbiota in mice supplemented with pomegranate peel extract [57]. The results showed a decrease in body weight and the insulin resistance index by 27.46% and 46.07%, respectively. The gut microbiota analysis indicated enrichment of *Bacteroides* with a decreased abundance of *Firmicutes* and *Proteobacteria* in pomegranate-supplemented mice. In addition, pomegranate also increased the abundance of *Akkermansia*, *Anaerotruncus*, *Lachnoclostridium*, and *Parabacteroides* species. In another study [58], pomegranate extract and Uro-A mitigated inflammatory responses and preserved colonic structure by modulating gut microbiota and downregulating the key inflammatory pathways in a rat model of colitis. Both pomegranate extract and Uro-A decreased inflammatory markers in colon mucosa, including inducible nitric oxide synthase (iNOS), prostaglandin E synthase, prostaglandin E, and cyclooxygenase-2. The fecal microbiota analysis indicated an increase in *Lactobacillus* and *Bifidobacterium* species in rats treated with pomegranate extract (250 mg/kg/day) and Uro-A (15 mg/kg/day) for 10 days, as compared to the control.

In recent years, there has been a notable increase in in vivo studies to investigate the biological effects of urolithins and confirm the health benefits associated with the consumption of ellagitannin (ET)-rich foods [59]. This growing interest is partly driven by the well-established fact that the bioavailability of ETs and EA is very low, primarily due to their poor absorption in the gastrointestinal tract. As a result, these compounds undergo extensive microbial metabolism in the colon, leading to the production of urolithins, which are thought to be the key bioactive metabolites responsible for the observed health benefits.

In our experimental conditions, Uro-A was detected at low concentrations exclusively in the fecal microbiota samples isolated from healthy adults following the in vitro simulated digestion and fermentation of the pomegranate extract. The most plausible explanation is that the fecal material of children and diseased subjects lacks sufficient microorganisms capable of converting EA into Uro-A. Moreover, according to Mena et al. [60], the digestion process can significantly affect the stability and availability of EA, potentially resulting in a considerable loss and reducing its availability for fermentation by gut microbiota. In fact, they demonstrated that gastric digestion significantly reduces the available amount of EA due to the highly acidic pH of the stomach, with a decrease from 45% to 5% during this phase. Although a slight recovery was observed in the intestinal phase, the EA available for fermentation was still reduced compared to the beginning of digestion, and extensive degradation also occurred for Uro-A and EA during the colonic step [60]. In our experiments, the simulated digestive phase may have degraded a significant amount of ETs, leaving an insufficient quantity for effective conversion into urolithins during fecal fermentation. This idea is further supported by the fact that many in vitro studies, such as those conducted by He et al. [61], have focused on assessing the capacity of the gut microbiota isolated from fecal material to metabolize ellagic acid into urolithins. In contrast, our findings differ from those of human studies, which have demonstrated that the intake of ellagitannin-rich foods, where digestion occurs naturally, enables the identification and quantification of the resulting metabolites in biological samples such as urine, plasma, and feces [62–64]. For instance, Mosele et al. showed that urolithins are produced from pomegranate extracts following in vitro digestion and fermentation, with considerable variability in the amount of urolithin derivatives due to differences in individual colonic microbiota [34]. Similarly, Tan et al. investigated the bioavailability of rambutan peel polyphenols (RPPs) and found that ellagic acid could be transformed into urolithins during in vitro gastrointestinal digestion, Caco-2 cell transport, and colonic fermentation, influenced by the action of intestinal microbiota [65].

In the present study, the simulated in vitro fermentation process ended after 24 h of incubation with human fecal material, but this may not have provided sufficient time for the complete conversion of EA into Uro-A, as reported by Garcia-Villalba et al., who observed that Uro-A is detected after 14 h and increases over time, stabilizing around 42 h during in vitro fecal fermentation of EA [63]. Similarly, Dacrema et al. demonstrated in vivo that while Uro-A levels in CD1 mouse plasma were relatively low after 24 h, they increased over time, reaching higher concentrations up to two weeks post-intake, indicating a delayed conversion process [66].

It is also known that the efficiency of EA conversion into urolithins can vary significantly between individuals, with differences in urolithin production profiles between "producers" and "non-producers," depending on the presence or absence of specific bacterial groups [67,68]. It is possible that the composition of the donors' fecal microbiota influenced the ability to convert EA into Uro-A, contributing to the absence of detectable metabolites. This may reflect the presence of microbial species that are not directly involved in the production of certain urolithins.

This study has its own limitations and strengths. The main limitation is the sole use of an in vitro experimental model, as unlike in vivo and clinical trials, in vitro experimental models are not true representatives of human systems. The simulated in vitro digestion and fermentation processes do not perfectly mimic what occurs in the human intestine. The in vitro conditions may not accurately reflect the complex and dynamic environment of the gut. First of all, the metabolic activity of gut microbes and urolithin production may vary depending on the availability of additional carbon sources, suggesting that future studies evaluate the impact of co-fermentation with other nutrients, such as prebiotics, for example, inulin, to better simulate real dietary conditions. Factors such as pH levels, enzyme activity, interactions with host tissues, and the presence of different microbial populations can significantly influence metabolic processes. Moreover, the time required for complete Uro-A production may vary across individuals and experiments, suggesting that longer incubation periods or alternative methodologies might be necessary to better assess urolithin production in in vitro experiments. Given the individual variability in the gut microbiota composition, future studies should also focus on characterizing the specific microbial taxa responsible for urolithin production and the conditions that optimize this metabolic conversion. In addition, the SCFA concentration in different fecal samples from the healthy and diseased subjects was determined, but the effect of the pomegranate extract on the diversity and activity of individual gut microbiota strains was not assessed. However, the impact of pomegranate extract on the functionality of the gut microbiota from healthy and diseased subjects (involving both adults and children) was determined for the first time, which is the major strength of the current investigation.

4. Materials and Methods

4.1. Chemicals and Reagents

Iron (III) chloride hexahydrate, 2,4,6-tri(2-pyridyl)-s-triazine (TPTZ), sodium acetate, 2,2-diphenyl-1-picrylhydrazyl hydrate (DPPH), 2,2′-azinobis(3-ethylbenzothiazoline)-6-sulfonic acid (ABTS), hydrochloric acid, methanol, 6-hydroxy-2,5,7,8-tetramethylchromane-2-carboxylic acid (Trolox), Folin–Ciocalteu reagent, sodium carbonate, gallic acid, potassium dihydrogen phosphate, potassium chloride, magnesium chloride hexahydrate, sodium chloride, calcium chloride dihydrate, sodium bicarbonate, ammonium carbonate, hydrochloric acid, salivary alpha-amylase, pepsin from porcine gastric mucosa, bile extract from porcine, sodium dihydrogen phosphate, sodium sulfide, tryptone, cysteine, resazurin, acetic acid, propionic acid, and butyric acid were obtained from Sigma-Aldrich (Germany). Analytical-grade methanol (MeOH), water, and formic acid (FA) were obtained from Carlo Erba Reagents (Milan, Italy). Urolithin A ($C_{13}H_8O_4$), with a purity of ≥90%, was acquired from Merck Group (Darmstadt, Germany) and used as a reference standard for quantitative analysis. Urolithin A was dissolved in methanol to produce a stock solution at a concentration of 1 mg/mL. By starting from an individual stock solution, the calibration curve was generated by diluting it in methanol containing 0.1% formic acid, resulting in a final concentration ranging from 0.01 to 5.0 mg/L.

The pomegranate fruit was macerated in water and chopped into small pieces. Subsequently, the extract was obtained by mixing a hydroalcoholic solution, specifically in a 70:30 ethanol/water ratio, in a single-phase extraction using ultrafiltration absorbent membranes (0.01–0.1 μm) to obtain a particle-free liquid extract. Finally, as reported by Ullah et al. [4], this extract was then subjected to a spray-drying process to obtain a dry extract.

4.2. In Vitro Digestion

The pomegranate extract underwent in vitro gastrointestinal digestion to simulate physiological human intestinal processes, following an adaptation of the protocol by Minekus et al. [69]. The digestion process consisted of three phases: oral, gastric, and intestinal. Initially, 5 g of the sample was mixed with 5 mL of simulated salivary fluid containing α-amylase (150 U/mL) and incubated at 37 °C for 5 min under continuous oscillation. Subsequently, 10 mL of simulated gastric fluid containing pepsin (4000 U/mL) were added, and the pH was adjusted to 3. The mixture was maintained at 37 °C for 2 h under oscillation. Finally, 20 mL of simulated intestinal fluid containing pancreatin (200 U/mL) and bile salts (20 mM) were introduced, and the pH was raised to 7. The sample was then incubated at 37 °C for another 2 h. To halt the enzymatic activity, the mixture was placed on ice for 15 min. The supernatant, representing the bioavailable fraction for small intestine absorption, was collected and stored at −20 °C for subsequent antioxidant capacity analysis. Meanwhile, the pellet, representing the undigested fraction destined for colonic fermentation, was directly used for in vitro fermentation. A control sample containing only enzymes and salts was also processed. All experiments were conducted in triplicate, generating three independent digestion liquids for subsequent fermentation.

4.3. In Vitro Fermentation

This process was conducted following the protocol described by Pérez-Burillo et al. [70], using fecal material from various Caucasian donors. Adult donors (aged 21–42 years) included healthy individuals (BMI, 18.5–24.99), obese individuals (BMI ≥ 30), and celiac individuals (BMI within the normal range, 18.5–24.99). Pediatric donors (aged 6–12 years) included healthy children (BMI, 5th–85th percentile), obese children (BMI > 95th percentile), and celiac children (BMI, 5th–85th percentile). Fecal material was collected using a sterile container (Fecotainer, AT Medical) and gloves, with the help of a disposable sterile spoon. Once collected, fecal material was stored by the volunteer at home in a refrigerator and transported to the laboratory in a cooler bag within 24 h. Upon arrival at the laboratory, fecal samples from three donors for each experimental condition were pooled to account for interindividual variability and used for in vitro fermentations. The study adhered to the guidelines of the Declaration of Helsinki and was approved by the Ethics Committee of the University of Granada (protocol code 1080/CEIH/2020) for the adult participants and by the Scientific Committee of the University Hospital "Attikon" (decision No. 546/1-10-2020) for the pediatric participants. None of the donors had taken antibiotics within the previous three months.

In vitro fermentation was performed in 15 mL centrifuge tubes. Each tube contained 500 mg of the solid digestion residue along with 10% of the final digestion volume. A control tube was also included, which did not contain the pomegranate extract. Additionally, each tube was supplemented with 7.5 mL of the fermentation growth medium (comprising 14 g/L of peptone, 312 mg/L of cysteine, 312 mg/L of hydrogen sulfide, and 0.1% *v/v* resazurin) and 2 mL of the fecal inoculum (prepared by suspending the fecal material in phosphate-buffered saline, 33% *w/v*).

To maintain anaerobic conditions, nitrogen gas was bubbled into the tubes, which were then incubated at 37 °C under oscillation for 20 h. Microbial activity was halted by placing the tubes on ice for 15 min. The samples were then centrifuged at 6000× *g* for 10 min, and the resulting supernatant was aliquoted and stored at −20 °C for subsequent antioxidant capacity and SCFA analysis. This process generated three fermentation liquids per experimental condition.

4.4. Antioxidant Assays

Both the supernatant from in vitro digestion and the supernatant from in vitro fermentation were analyzed for their antioxidant capacity. The total antioxidant capacity was determined as the sum of the antioxidant capacities of these two fractions. To ensure accuracy, respective blanks (containing only chemicals, enzymes, and the inoculum) were used to correct the antioxidant capacity values for each method. To evaluate the impact of digestion and fermentation on the TPC and antioxidant capacity of the pomegranate extract, the digested samples were compared with the non-digested samples, while the fermented samples (from healthy subjects) were compared with the non-fermented (digested) samples.

All experiments were performed in triplicate for both digestion and fermentation liquids.

4.4.1. Total Phenolic Content (TPC)

The TPC of the pomegranate samples was determined using the Folin–Ciocalteu assay, following the protocol of Moreno-Montoro et al. [71]. An aliquot of 30 μL of the digestion or fermentation supernatant was added in triplicate to a 96-well plate (Cytation, Agilent Technologies Inc., Santa Clara, CA, USA) and mixed with 60 μL of 10% (*w/v*) sodium carbonate, 195 μL of ultra-pure water, and 15 μL of the Folin–Ciocalteu reagent. The antioxidant reaction was monitored for 60 min at 37 °C. A calibration curve was prepared using gallic acid at concentrations ranging from 0.01 to 1.00 mg/mL. The results were expressed as mg gallic acid equivalent per gram (mg GAE/g) of pomegranate extract.

4.4.2. DPPH Assay

The DPPH assay was performed following the protocol of Yen and Chen [72]. Twenty μL of the digestion or fermentation supernatant were mixed with 280 μL of a freshly prepared DPPH reagent in a 96-well plate. The DPPH reagent was prepared daily by dissolving 74 mg of DPPH in 1 L of methanol. The reaction was monitored for one hour at 37 °C using a Cytation 5 plate reader (Agilent Technologies Inc., Santa Clara, CA, USA). Each sample was analyzed in triplicate. A calibration curve was prepared using Trolox (0.01–0.4 mg/mL), and the results were expressed as mg Trolox equivalent per gram of pomegranate extract.

4.4.3. ABTS Assay

The ABTS+ assay was performed following the method described by Re et al. [73], with slight modifications. The ABTS+ solution was prepared by reacting a 7 mM ABTS+ stock solution with 2.45 mM potassium persulfate and allowing the mixture to stand in the dark at room temperature for 12–16 h before use. Before the assay, the ABTS+ solution was diluted with 5 mM phosphate-buffered saline (pH 7.4) to achieve an absorbance of 0.70 ± 0.02 at 730 nm. Then, 10 μL of the sample were added to 4 mL of the diluted ABTS+ solution, and absorbance was measured after 20 min using a Cytation 5 plate reader (Agilent Technologies Inc., Santa Clara, CA, USA). A calibration curve was prepared using the Trolox stock solution, and the results were expressed as mg Trolox equivalent per gram of sample.

4.4.4. FRAP Assay

The FRAP assay was performed following the method of Benzie and Strain [74] using a Cytation 5 microplate reader (Agilent Technologies Inc., Santa Clara, CA, USA). For each sample, including the digestion and fermentation supernatants, 20 μL were mixed with 280 μL of a freshly prepared FRAP reagent in a 96-well plate. The FRAP reagent consisted of 0.3 M sodium acetate buffer (pH 3.6), ferric chloride solution, and 40 mM 2,4,6-tri(2-pyridyl)-s-triazine (TPTZ) solution in a ratio of 10:1:1. The reaction was conducted at 37 °C, and absorbance was measured at 595 nm every 30 s for 30 min. Each sample was analyzed in triplicate to ensure accuracy. A calibration curve was prepared using Trolox at concentrations ranging from 0.01 to 0.4 mg/mL, and the results were expressed as mg Trolox equivalent per gram of pomegranate extract.

4.5. Analysis of Short-Chain Fatty Acids (SCFAs)

The preparation of samples for chromatographic analysis involved centrifuging the fermentation supernatant at 13,300 rpm for 5 min. The supernatant was then filtered through a 0.22 μm filter and diluted in a 1:10 ratio with 1 M hydrochloric acid, following the protocol described by Panzella et al. [75]. The analysis was performed using an Agilent Poroshell 120 SB-Aq column (3 × 150 mm, 2.7 μm) with 5 mM sulfuric acid as the mobile phase under isocratic elution at a flow rate of 0.5 mL/min. A sample injection volume of 5 μL was used, and both the column and refractive index detector (RID) temperatures were maintained at 35 °C. Chromatographic analysis was conducted in triplicate for each sample using an Agilent 1290 Infinity II system (Agilent Technologies Inc., Santa Clara, CA, USA).

For quantification, calibration was performed using external standards of lactic, acetic, succinic, propionic, *n*-butyric, and isobutyric acids. The *n*-butyric and isobutyric acids were quantified together. The total SCFAs were calculated as the sum of acetic, propionic, and butyric acids, and the results were expressed in mmol of each acid per liter of the fermentation supernatant. The SCFA analysis was carried out in triplicate for each experimental condition, including healthy, obese, and celiac subjects from both adult and pediatric groups.

4.6. Extraction of Urolithin A

The extraction of urolithin A was carried out following the protocol described by García-Villalba et al. [76], with some modifications. Five milligrams of each freeze-dried sample were placed in a 50 mL tube, and 5 mL of methanol acidified with 0.1% formic acid were added. The mixture was vortexed for 3 min using a MIX ARGOlab Vortex Mixer GB22008013. Following this, the sample was sonicated for 10 min in an ultrasonic bath (LBS 1-6 Lt; FALC, Treviglio (BG), Italy) and then centrifuged at 5000 rpm for 5 min at 4 °C using an SL16R centrifuge (Thermo Fisher Scientific, Dreieich, Germany). The supernatant was collected and filtered through a 0.2 μm nylon syringe filter. Finally, ultrahigh-performance liquid chromatography coupled with high-resolution mass spectrometry (UHPLC Q-Orbitrap HRMS; Thermo Fisher Scientific, Dreieich, Germany) was used for the analysis.

4.7. UHPLC Q-Orbitrap HRMS Analysis of Urolithin A

The qualitative and quantitative measurement of urolithin A was conducted using ultrahigh-performance liquid chromatography coupled with high-resolution mass spectrometry (UHPLC Q-Orbitrap HRMS). Mass spectrometry analysis was performed on a Q-Exactive mass spectrometer (Thermo Fisher Scientific, Waltham, MA, USA) equipped with an electrospray ionization (ESI) source. The analysis was carried out in the negative ionization mode, targeting a mass range of 80 to 500 m/z. Data acquisition was performed in the full-ion mass spectrometry (MS) mode, with the following key parameters: scan range of 80–500 m/z, resolution of 70,000 full width at half maximum (FWHM), microscan set to 1, automatic gain control (AGC) target of 1×10^6, maximum injection time of 200 ms, sheath gas flow rate of 35, auxiliary gas flow rate of 10, spray voltage of 2.8 kV, capillary temperature of 375 °C, S-lens RF level of 50, and auxiliary gas heater temperature of 350 °C. A mass tolerance of 5 ppm was maintained for the identification and confirmation of urolithin A ions.

Chromatographic separation was performed using a UHPLC system (Dionex UltiMate 3000; Thermo Fisher Scientific, Waltham, MA, USA) equipped with a degassing unit, a quaternary pump, and an autosampler. A Kinetex F5 column (2.6 µm particle size, 100 mm × 2.1 mm) from Phenomenex (Torrance, CA, USA) was used, maintained at 20 °C. The mobile phase consisted of water (A) and methanol (B), both acidified with 0.1% formic acid. Optimal separation was achieved using the following gradient elution program: from 0 to 0.5 min at 100% A, followed by a decrease from 100% to 30% A between 0.5 and 1 min, and then further decreasing from 30% to 0% A from 1 to 7.8 min. The mobile phase was held at 0% A from 7.8 to 8.3 min, followed by re-equilibration back to 100% A from 8.3 to 10 min. The flow rate was set at 400 µL/min, with an injection volume of 5 µL. Data analysis was performed using Xcalibur software (version 3.1.66.19; Thermo Fisher Scientific).

4.8. Statistical Analysis

For the statistical analysis, the data were presented as the means ± standard deviation (SD). A paired t-test was utilized to determine the differences in the pomegranate extract before and after in vitro oral, gastric, and intestinal digestion processes. Additionally, statistical comparisons between the groups were conducted using one-way ANOVA followed by Dunnett's post-hoc test for multiple comparisons to determine significance, which was set at $p < 0.05$. Concerning the antioxidant profile of the pomegranate extract, two different analyses were performed to assess the differences between the digested and fermented samples, and the differences between the healthy subjects (control group), obese, and celiac individuals after in vitro fermentation. The statistical analyses were performed using GraphPad Prism, version 8 (San Diego, CA, USA). Finally, the statistical analysis regarding the SCFAs was conducted using the SPSS statistical program. The mean values and standard deviations for each group were determined, and the statistical significance of the data was assessed using the non-parametric Kolmogorov–Smirnov analysis of variance test, followed by Bonferroni post-hoc tests to compare the samples that exhibited significant variation ($p < 0.05$).

5. Conclusions

This study highlights the antioxidant potential and gut microbiota-modulating effects of fermented pomegranate extract, demonstrating its capacity to enhance SCFA production, particularly lactic acid, in individuals with obesity and celiac disease. Given the critical role of gut microbiota in metabolic and gastrointestinal disorders, dietary interventions targeting microbial composition and activity represent a promising strategy for disease prevention and management. Notably, polyphenols and the gut microbiota share a bidirectional relationship: polyphenols influence microbial diversity and activity, while the gut microbiota transforms polyphenols into bioactive metabolites such as urolithins, enhancing their health benefits. Our findings support the potential prebiotic application of pomegranate extract in functional foods and dietary supplements, offering a natural approach to modulating the gut microbiota in at-risk populations.

However, this study is limited by the absence of next-generation sequencing (NGS) data, which would provide a deeper understanding of microbial shifts and specific bacterial taxa involved in polyphenol metabolism. Future research should integrate metagenomic and metabolomic approaches to further elucidate the mechanistic interactions between pomegranate polyphenols and the gut microbiota. Such insights could facilitate the development of personalized nutrition strategies and targeted functional foods for improved gut health and disease management.

Author Contributions: Conceptualization, D.G.B., J.Á.R.-H. and M.D.; investigation, D.G.B., A.D.-O. and J.Á.R.-H.; data curation, D.G.B. and A.D.-O.; methodology, A.D.-O. and L.I.; writing—original draft preparation, D.G.B., H.U. and M.D.; formal analysis, A.D.-O. and S.L.; visualization, D.G.B.; writing—review and editing, J.Á.R.-H., L.F.D.L., M.V.M., D.M., A.D.M. and C.V.R.; project administration, C.V.R.; supervision, J.Á.R.-H. and M.D.; funding acquisition, J.Á.R.-H. All authors have read and agreed to the published version of the manuscript.

Funding: This research was supported by the "Plan propio de Investigación y Transferencia" of the University of Granada under the program "Intensificación de la Investigación, modalidad B", granted to José Á. Rufián-Henares.

Institutional Review Board Statement: The study was conducted in accordance with the Declaration of Helsinki, and was approved by the Ethics Committee of the University of Granada (protocol code 1080/CEIH/2020) for the adult participants and by the Scientific Committee of the University Hospital "Attikon" (decision No. 546/1-10-2020) for the pediatric participants.

Informed Consent Statement: Informed consent was obtained from all subjects involved in the study.

Data Availability Statement: The original contributions presented in the study are included in the article, further inquiries can be directed to the corresponding author.

Acknowledgments: The authors would like to thank ESSERRE Pharma srl for providing the pomegranate samples tested in this investigation.

Conflicts of Interest: C.V.R. is an employee of ESSERRE Pharma srl. None of the academic researchers listed as co-authors served as consultants for ESSERRE srl or received any personal compensation from it.

Abbreviations

The following abbreviations are used in this manuscript:

ABTS	2,2′-azinobis(3-ethylbenzothiazoline)-6-sulfonic acid
AP-1	Activator protein
COX	Cyclooxygenase
DPPH	2,2-diphenyl-1-picrylhydrazyl hydrate
EA	Ellagic acid
H_2O_2	Hydrogen peroxide
IL	Interleukins
NF-κB	Nuclear factor kappa B
PPARs	Peroxisome proliferator-activated receptors
SCFAs	Short-chain fatty acids

STAT	Signal transducer and activator of transcription
TNF-α	Tumor necrosis factor
TPC	Total phenolic content
TPTZ	2,4,6-tri(2-pyridyl)-s-triazine
Trolox	6-hydroxy-2,5,7,8-tetramethylchromane-2-carboxylic acid
UHPLC Q-Orbitrap HRMS	Ultrahigh-performance liquid chromatography coupled with high-resolution mass spectrometry
Uro-A	Urolithin A

References

1. Celik, I.; Temur, A.; Isik, I. Hepatoprotective role and antioxidant capacity of pomegranate (Punica granatum) flowers infusion against trichloroacetic acid-exposed in rats. *Food Chem. Toxicol.* **2009**, *47*, 145–149. [CrossRef]
2. Lansky, E.P.; Newman, R.A. Punica granatum (pomegranate) and its potential for prevention and treatment of inflammation and cancer. *J. Ethnopharmacol.* **2007**, *109*, 177–206. [CrossRef] [PubMed]
3. USDA. Pomegranates, Raw. Available online: https://fdc.nal.usda.gov/food-details/169134/nutrients (accessed on 25 March 2025).
4. Ullah, H.; Sommella, E.; Minno, A.D.; Piccinocchi, R.; Buccato, D.G.; De Lellis, L.F.; Riccioni, C.; Baldi, A.; El-Seedi, H.R.; Khalifa, S.A.; et al. Combination of chemically characterized pomegranate extract and hydrophilic vitamins against prolonged fatigue: A monocentric, randomized, double-blind, placebo-controlled clinical trial. *Nutrients* **2023**, *15*, 2883. [CrossRef]
5. Coronado-Reyes, J.A.; Cortes-penagos, C.D.J.; González-Hernández, J.C. Chemical composition and great applications to the fruit of the pomegranate (*Punica granatum*): A review. *Food Sci. Technol.* **2021**, *42*, e29420. [CrossRef]
6. Ismail, T.; Sestili, P.; Akhtar, S. Pomegranate peel and fruit extracts: A review of potential anti-inflammatory and anti-infective effects. *J. Ethnopharmacol.* **2012**, *143*, 397–405. [CrossRef] [PubMed]
7. Hussain, T.; Tan, B.; Yin, Y.; Blachier, F.; Tossou, M.C.B.; Rahu, N. Oxidative stress and inflammation: What polyphenols can do for us? *Oxid. Med. Cell. Longev.* **2016**, *2016*, 7432797. [CrossRef]
8. Rodrigo, R.; Miranda, A.; Vergara, L. Modulation of endogenous antioxidant system by wine polyphenols in human disease. *Clin. Chim. Acta* **2011**, *412*, 410–424. [CrossRef] [PubMed]
9. Tsai, H.-Y.; Ho, C.-T.; Chen, Y.-K. Biological actions and molecular effects of resveratrol, pterostilbene, and 3′-hydroxypterostilbene. *Yao Wu Shi Pin Fen Xi* **2017**, *25*, 134–147. [CrossRef]
10. Fraga, C.G.; Galleano, M.; Verstraeten, S.V.; Oteiza, P.I. Basic biochemical mechanisms behind the health benefits of polyphenols. *Mol. Aspects Med.* **2010**, *31*, 435–445. [CrossRef]
11. Aloqbi, A.; Omar, U.; Yousr, M.; Grace, M.; Lila, M.A.; Howell, N. Antioxidant activity of pomegranate juice and punicalagin. *Nat. Sci.* **2016**, *8*, 235–246. [CrossRef]
12. Seeram, N.P.; Aviram, M.; Zhang, Y.; Henning, S.M.; Feng, L.; Dreher, M.; Heber, D. Comparison of antioxidant potency of commonly consumed polyphenol-rich beverages in the United States. *J. Agric. Food Chem.* **2008**, *56*, 1415–1422. [CrossRef]
13. Vučić, V.; Grabež, M.; Trchounian, A.; Arsić, A. Composition and potential health benefits of pomegranate: A review. *Curr. Pharm. Des.* **2019**, *25*, 1817–1827. [PubMed]
14. Bialonska, D.; Ramnani, P.; Kasimsetty, S.G.; Muntha, K.R.; Gibson, G.R.; Ferreira, D. The influence of pomegranate by-product and punicalagins on selected groups of human intestinal microbiota. *Int. J. Food Microbiol.* **2010**, *140*, 175–182. [CrossRef] [PubMed]
15. Bialonska, D.; Kasimsetty, S.G.; Schrader, K.K.; Ferreira, D. The effect of pomegranate (Punica granatum L.) byproducts and ellagitannins on the growth of human gut bacteria. *J. Agric. Food Chem.* **2009**, *57*, 8344–8349. [CrossRef]
16. Tan, J.; McKenzie, C.; Potamitis, M.; Thorburn, A.N.; Mackay, C.R.; Macia, L. The role of short-chain fatty acids in health and disease. *Adv. Immunol.* **2014**, *121*, 91–119. [PubMed]
17. Huang, W.; Guo, H.L.; Deng, X.; Zhu, T.T.; Xiong, J.F.; Xu, Y.H.; Xu, Y. Short-chain fatty acids inhibit oxidative stress and inflammation in mesangial cells induced by high glucose and lipopolysaccharide. *Exp. Clin. Endocrinol. Diabetes* **2017**, *125*, 98–105. [CrossRef]
18. Den Besten, G.; Bleeker, A.; Gerding, A.; van Eunen, K.; Havinga, R.; van Dijk, T.H.; Oosterveer, M.H.; Jonker, J.W.; Groen, A.K.; Reijngoud, D.J.; et al. Short-chain fatty acids protect against high-fat diet–induced obesity via a PPARγ-dependent switch from lipogenesis to fat oxidation. *Diabetes* **2015**, *64*, 2398–2408. [CrossRef]
19. Mokdad, A.H.; Bowman, B.A.; Ford, E.S.; Vinicor, F.; Marks, J.S.; Koplan, J.P. The continuing epidemics of obesity and diabetes in the United States. *JAMA* **2001**, *286*, 1195–1200. [CrossRef]

20. Desvergne, B.; Wahli, W. Peroxisome proliferator-activated receptors: Nuclear control of metabolism. *Endocr. Rev.* **1999**, *20*, 649–688.
21. Ricote, M.; Li, A.C.; Willson, T.M.; Kelly, C.J.; Glass, C.K. The peroxisome proliferator-activated receptor-gamma is a negative regulator of macrophage activation. *Nature* **1998**, *391*, 79–82. [CrossRef]
22. Jiménez-Zamora, A.; Delgado-Andrade, C.; Rufián-Henares, J.A. Antioxidant capacity, total phenols and color profile during the storage of selected plants used for infusion. *Food Chem.* **2016**, *199*, 339–346. [PubMed]
23. Derakhshan, Z.; Ferrante, M.; Tadi, M.; Ansari, F.; Heydari, A.; Hosseini, M.S.; Conti, G.O.; Sadrabad, E.K. Antioxidant activity and total phenolic content of ethanolic extract of pomegranate peels, juice and seeds. *Food Chem. Toxicol.* **2018**, *114*, 108–111. [PubMed]
24. Ardekani, M.R.S.; Hajimahmoodi, M.; Oveisi, M.R.; Sadeghi, N.; Jannat, B.; Ranjbar, A.M.; Gholam, N.; Moridi, T. Comparative antioxidant activity and total flavonoid content of Persian pomegranate (*Punica granatum* L.) cultivars. *Iranian J. Pharm. Res.* **2011**, *10*, 519.
25. Fawole, O.A.; Opara, U.L. Stability of total phenolic concentration and antioxidant capacity of extracts from pomegranate co-products subjected to in vitro digestion. *BMC Complement. Altern. Med.* **2016**, *16*, 358.
26. Chen, G.L.; Chen, S.G.; Zhao, Y.Y.; Luo, C.X.; Li, J.; Gao, Y.Q. Total phenolic contents of 33 fruits and their antioxidant capacities before and after in vitro digestion. *Ind. Crops Prod.* **2014**, *57*, 150–157. [CrossRef]
27. Chiang, C.J.; Kadouh, H.; Zhou, K. Phenolic compounds and antioxidant properties of gooseberry as affected by in vitro digestion. *LWT-Food Sci. Technol.* **2013**, *51*, 417–422.
28. Pavan, V.; Sancho, R.A.S.; Pastore, G.M. The effect of in vitro digestion on the antioxidant activity of fruit extracts (*Carica papaya*, *Artocarpus heterophillus* and *Annona marcgravii*). *LWT-Food Sci. Technol.* **2014**, *59*, 1247–1251.
29. Bouayed, J.; Hoffmann, L.; Bohn, T. Total phenolics, flavonoids, anthocyanins and antioxidant activity following simulated gastro-intestinal digestion and dialysis of apple varieties: Bioaccessibility and potential uptake. *Food Chem.* **2011**, *128*, 14–21. [CrossRef]
30. Arranz, S.; Silván, J.M.; Saura-Calixto, F. Nonextractable polyphenols, usually ignored, are the major part of dietary polyphenols: A study on the Spanish diet. *Mol. Nutr. Food Res.* **2010**, *54*, 1646–1658.
31. Frontela-Saseta, C.; López-Nicolás, R.; González-Bermúdez, C.A.; Peso-Echarri, P.; Ros-Berruezo, G.; Martínez-Graciá, C.; Canali, R.; Virgili, F. Evaluation of antioxidant activity and antiproliferative effect of fruit juices enriched with Pycnogenol® in colon carcinoma cells. The effect of in vitro gastrointestinal digestion. *Phytother. Res.* **2011**, *25*, 1870–1875. [CrossRef]
32. Gullon, B.; Pintado, M.E.; Fernández-López, J.; Pérez-Álvarez, J.A.; Viuda-Martos, M. In vitro gastrointestinal digestion of pomegranate peel (Punica granatum) flour obtained from co-products: Changes in the antioxidant potential and bioactive compounds stability. *J. Funct. Foods* **2015**, *19*, 617–628. [CrossRef]
33. Pérez-Vicente, A.; Gil-Izquierdo, A.; García-Viguera, C. In vitro gastrointestinal digestion study of pomegranate juice phenolic compounds, anthocyanins, and vitamin C. *J. Agric. Food Chem.* **2002**, *50*, 2308–2312. [PubMed]
34. Mosele, J.I.; Macià, A.; Romero, M.P.; Motilva, M.J.; Rubió, L. Application of in vitro gastrointestinal digestion and colonic fermentation models to pomegranate products (juice, pulp and peel extract) to study the stability and catabolism of phenolic compounds. *J. Funct. Foods* **2015**, *14*, 529–540.
35. De Pinedo, A.T.; Peñalver, P.; Morales, J.C. Synthesis and evaluation of new phenolic-based antioxidants: Structure–activity relationship. *Food Chem.* **2007**, *103*, 55–61. [CrossRef]
36. Popovici, C.; Saykova, I.; Tylkowski, B. Evaluation of the antioxidant activity of phenolic compounds through reactivity with the free radical DPPH. *Rev. Genie Ind.* **2010**, *4*, 131–887.
37. Yang, F.; Chen, C.; Ni, D.; Yang, Y.; Tian, J.; Li, Y.; Chen, S.; Ye, X.; Wang, L. Effects of fermentation on bioactivity and the composition of polyphenols contained in polyphenol-rich foods: A review. *Foods* **2023**, *12*, 3315. [CrossRef] [PubMed]
38. Díaz-Mula, H.M.; Tomás-Barberán, F.A.; García-Villalba, R. Pomegranate fruit and juice (cv. Mollar), rich in ellagitannins and anthocyanins, also provide a significant content of a wide range of proanthocyanidins. *J. Agric. Food Chem.* **2019**, *67*, 9160–9167.
39. Liu, Q.; Hua, Z.; Chen, M.; Liu, S.; Ahmed, S.; Hou, X.; Yang, G.; Fang, Y. Changes in polyphenols and antioxidant properties of pomegranate peels fermented by urolithin A-producing *Streptococcus thermophilus* FUA329. *ACS Food Sci. Technol.* **2023**, *3*, 1383–1392.
40. Singh, V.; Ahlawat, S.; Mohan, H.; Gill, S.S.; Sharma, K.K. Balancing reactive oxygen species generation by rebooting gut microbiota. *J. Appl. Microbiol.* **2022**, *132*, 4112–4129.
41. De Filippis, A.; Ullah, H.; Baldi, A.; Dacrema, M.; Esposito, C.; Garzarella, E.U.; Santarcangelo, C.; Tantipongpiradet, A.; Daglia, M. Gastrointestinal disorders and metabolic syndrome: Dysbiosis as a key link and common bioactive dietary components useful for their treatment. *Int. J. Mol. Sci.* **2020**, *21*, 4929. [CrossRef]
42. Barbara, G.; Feinle-Bisset, C.; Ghoshal, U.C.; Santos, J.; Vanner, S.J.; Vergnolle, N.; Zoetendal, E.G.; Quigley, E.M. The intestinal microenvironment and functional gastrointestinal disorders. *Gastroenterology* **2016**, *150*, 1305–1318. [CrossRef] [PubMed]

43. Plamada, D.; Vodnar, D.C. Polyphenols—Gut microbiota interrelationship: A transition to a new generation of prebiotic *Nutrients* **2021**, *14*, 137. [CrossRef] [PubMed]
44. Tang, T.W.; Chen, H.C.; Chen, C.Y.; Yen, C.Y.; Lin, C.J.; Prajnamitra, R.P.; Chen, L.L.; Ruan, S.C.; Lin, J.H.; Lin, P.J.; et al. Los of gut microbiota alters immune system composition and cripples postinfarction cardiac repair. *Circulation* **2019**, *139*, 647–65 [CrossRef] [PubMed]
45. Ecklu-Mensah, G.; Choo-Kang, C.; Maseng, M.G.; Donato, S.; Bovet, P.; Viswanathan, B.; Bedu-Addo, K.; Plange-Rhule, J.; O Boateng, P.; Forrester, T.E.; et al. Gut microbiota and fecal short chain fatty acids differ with adiposity and country of origin: Th METS-microbiome study. *Nat. Commun.* **2023**, *14*, 5160. [CrossRef]
46. Vinolo, M.A.; Rodrigues, H.G.; Nachbar, R.T.; Curi, R. Regulation of inflammation by short chain fatty acids. *Nutrients* **2011**, *3* 858–876. [CrossRef]
47. Gao, Z.; Yin, J.; Zhang, J.; Ward, R.E.; Martin, R.J.; Lefevre, M.; Cefalu, W.T.; Ye, J. Butyrate improves insulin sensitivity and increases energy expenditure in mice. *Diabetes* **2009**, *58*, 1509–1517. [CrossRef]
48. Henagan, T.M.; Stefanska, B.; Fang, Z.; Navard, A.M.; Ye, J.; Lenard, N.R.; Devarshi, P.P. Sodium butyrate epigenetically modulates high-fat diet-induced skeletal muscle mitochondrial adaptation, obesity and insulin resistance through nucleosome positioning *Br. J. Pharmacol.* **2015**, *172*, 2782–2798. [CrossRef]
49. Lu, Y.; Fan, C.; Li, P.; Lu, Y.; Chang, X.; Qi, K. Short chain fatty acids prevent high-fat-diet-induced obesity in mice by regulating G protein-coupled receptors and gut microbiota. *Sci. Rep.* **2016**, *6*, 37589. [CrossRef]
50. Bonomo, R.R.; Cook, T.M.; Gavini, C.K.; White, C.R.; Jones, J.R.; Bovo, E.; Zima, A.V.; Brown, I.A.; Dugas, L.R.; Zakharian, E.; et al Fecal transplantation and butyrate improve neuropathic pain, modify immune cell profile, and gene expression in the PNS of obese mice. *Proc. Natl. Acad. Sci. USA* **2020**, *117*, 26482–26493. [CrossRef]
51. Nistal, E.; Caminero, A.; Vivas, S.; de Morales, J.M.R.; de Miera, L.E.S.; Rodríguez-Aparicio, L.B.; Casqueiro, J. Differences in faecal bacteria populations and faecal bacteria metabolism in healthy adults and celiac disease patients. *Biochimie* **2012**, *94*, 1724–1729. [CrossRef]
52. Di Cagno, R.; De Angelis, M.; De Pasquale, I.; Ndagijimana, M.; Vernocchi, P.; Ricciuti, P.; Gagliardi, F.; Laghi, L.; Crecchio, C.; Guerzoni, M.E.; et al. Duodenal and faecal microbiota of celiac children: Molecular, phenotype and metabolome characterization. *BMC Microbiol.* **2011**, *11*, 219.
53. Di Cagno, R.; Rizzello, C.G.; Gagliardi, F.; Ricciuti, P.; Ndagijimana, M.; Francavilla, R.; Guerzoni, M.E.; Crecchio, C.; Gobbetti, M.; De Angelis, M. Different fecal microbiotas and volatile organic compounds in treated and untreated children with celiac disease. *Appl. Environ. Microbiol.* **2009**, *75*, 3963–3971.
54. Lerner, A.; Patricia, J.; Matthias, T. Nutrients, bugs and us: The short-chain fatty acids story in celiac disease. *Int. J. Celiac Dis.* **2016**, *4*, 92–94.
55. Abedi, E.; Hashemi, S.M.B. Lactic acid production–producing microorganisms and substrates sources-state of art. *Heliyon* **2020**, *6*, e04974. [PubMed]
56. Mosele, J.I.; Gosalbes, M.J.; Macia, A.; Rubio, L.; Vázquez-Castellanos, J.F.; Jimenez Hernandez, N.; Moya, A.; Latorre, A.; Motilva, M.J. Effect of daily intake of pomegranate juice on fecal microbiota and feces metabolites from healthy volunteers. *Mol. Nutr. Food Res.* **2015**, *59*, 1942–1953. [PubMed]
57. Song, H.; Shen, X.; Deng, R.; Chu, Q.; Zheng, X. Pomegranate peel anthocyanins prevent diet-induced obesity and insulin resistance in association with modulation of the gut microbiota in mice. *Eur. J. Nutr.* **2022**, *61*, 1837–1847.
58. Larrosa, M.; González-Sarrías, A.; Yáñez-Gascón, M.J.; Selma, M.V.; Azorín-Ortuño, M.; Toti, S.; Tomás-Barberán, F.; Dolara, P.; Espín, J.C. Anti-inflammatory properties of a pomegranate extract and its metabolite urolithin-A in a colitis rat model and the effect of colon inflammation on phenolic metabolism. *J. Nutr. Biochem.* **2010**, *21*, 717–725.
59. García-Villalba, R.; Giménez-Bastida, J.A.; Cortés-Martín, A.; Ávila-Gálvez, M.Á.; Tomás-Barberán, F.A.; Selma, M.V.; Espín, J.C.; González-Sarrías, A. Urolithins: A comprehensive update on their metabolism, bioactivity, and associated gut microbiota. *Mol. Nutr. Food Res.* **2022**, *66*, e2101019.
60. Mena, P.; Dall'Asta, M.; Calani, L.; Brighenti, F.; Del Rio, D. Gastrointestinal stability of urolithins: An in vitro approach. *Eur. J. Nutr.* **2015**, *56*, 99–106.
61. He, F.; Bian, Y.; Zhao, Y.; Xia, M.; Liu, S.; Gui, J.; Hou, X.; Fang, Y. In vitro conversion of ellagic acid to urolithin A by different gut microbiota of urolithin metabotype A. *Appl. Microbiol. Biotechnol.* **2024**, *108*, 215.
62. González-Barrio, R.; Edwards, C.A.; Crozier, A. Colonic catabolism of ellagitannins, ellagic acid, and raspberry anthocyanins: In vivo and in vitro studies. *Drug Metab. Dispos.* **2011**, *39*, 1680–1688. [CrossRef] [PubMed]
63. García-Villalba, R.; Beltrán, D.; Espín, J.C.; Selma, M.V.; Tomás-Barberán, F.A. Time course production of urolithins from ellagic acid by human gut microbiota. *J. Agric. Food Chem.* **2013**, *61*, 8797–8806. [CrossRef] [PubMed]
64. Cerdá, B.; Periago, P.; Espín, J.C.; Tomás-Barberán, F.A. Identification of urolithin a as a metabolite produced by human colon microflora from ellagic acid and related compounds. *J. Agric. Food Chem.* **2005**, *53*, 5571–5576. [CrossRef]

65. Tan, J.; Ma, Q.; Li, J.; Liu, Q.; Zhuang, Y. Bioavailability and antioxidant activity of Rambutan (Nephelium lappaceum) peel polyphenols during in vitro simulated gastrointestinal digestion, Caco-2 monolayer cell model application, and colonic fermentation. *J. Agric. Food Chem.* **2023**, *71*, 15829–15841. [CrossRef]
66. Dacrema, M.; Sommella, E.; Santarcangelo, C.; Bruno, B.; Marano, M.G.; Insolia, V.; Saviano, A.; Campiglia, P.; Stornaiuolo, M.; Daglia, M. Metabolic profiling, in vitro bioaccessibility and in vivo bioavailability of a commercial bioactive Epilobium angustifolium L. extract. *Biomed. Pharmacother.* **2020**, *131*, 110670. [CrossRef]
67. Tomás-Barberán, F.A.; González-Sarrías, A.; García-Villalba, R.; Núñez-Sánchez, M.A.; Selma, M.V.; García-Conesa, M.T.; Espín, J.C. Urolithins, the rescue of "old" metabolites to understand a "new" concept: Metabotypes as a nexus among phenolic metabolism, microbiota dysbiosis, and host health status. *Mol. Nutr. Food Res.* **2016**, *61*, 1500901. [CrossRef] [PubMed]
68. Iglesias-Aguirre, C.; Villalba, R.; Beltrán Riquelme, D.; Frutos Lisón, M.; Espín, J.C.; Tomás-Barberán, F.; Selma, M. Gut bacteria involved in ellagic acid metabolism to yield human urolithin metabotypes revealed. *J. Agric. Food Chem.* **2023**, *71*, 4029–4035. [CrossRef] [PubMed]
69. Minekus, M.; Alminger, M.; Alvito, P.; Balance, S.; Bohn, T.; Bourlieu, C.; Carrière, F.; Boutrou, R.; Corredig, M.; Dupont, D.; et al. A standardised static in vitro digestion method suitable for food–an international consensus. *Food Funct.* **2014**, *5*, 1113–1124. [CrossRef]
70. Pérez-Burillo, S.; Molino, S.; Navajas-Porras, B.; Valverde-Moya, Á.J.; Hinojosa-Nogueira, D.; López-Maldonado, A.; Pastoriza, S.; Rufián-Henares, J.Á. An in vitro batch fermentation protocol for studying the contribution of food to gut microbiota composition and functionality. *Nat. Protoc.* **2021**, *16*, 3186–3209. [CrossRef]
71. Moreno-Montoro, M.; Olalla-Herrera, M.; Gimenez-Martinez, R.; Navarro-Alarcon, M.; Rufián-Henares, J.A. Phenolic compounds and antioxidant activity of Spanish commercial grape juices. *J. Food Compos. Anal.* **2015**, *38*, 19–26.
72. Yen, G.; Chen, H. Antioxidant activity of various tea extracts in relation to their antimutagenicity. *J. Agric. Food Chem.* **1995**, *43*, 27–32. [CrossRef]
73. Re, R.; Pellegrini, N.; Proteggente, A.; Pannala, A.; Yang, M.; Rice-Evans, C. Antioxidant activity applying an improved ABTS radical cation decolorization assay. *Free Radic. Biol. Med.* **1999**, *26*, 1231–1237. [CrossRef] [PubMed]
74. Benzie, I.F.; Strain, J.J. The ferric reducing ability of plasma (FRAP) as a measure of "antioxidant power": The FRAP assay. *Anal. Biochem.* **1996**, *239*, 70–76. [CrossRef]
75. Panzella, L.; Pérez-Burillo, S.; Pastoriza, S.; Martín, M.Á.; Cerruti, P.; Goya, L.; Ramos, S.; Rufián-Henares, J.Á.; Napolitano, A.; d'Ischia, M. High Antioxidant action and prebiotic activity of hydrolyzed spent coffee grounds (HSCG) in a simulated digestion–fermentation model: Toward the development of a novel food supplement. *J. Agric. Food Chem.* **2017**, *65*, 6452–6459. [CrossRef]
76. García-Villalba, R.; Selma, M.V.; Espín, J.C.; Tomás-Barberán, F.A. Identification of novel urolithin metabolites in human feces and urine after the intake of a pomegranate extract. *J. Agric. Food Chem.* **2019**, *67*, 11099–11107. [CrossRef] [PubMed]

molecules

MDPI

Article

Development of a Scalable Extraction Process for Anthocyanins of Haskap Berry (*Lonicera caerulea*)

Damith Costa and H.P. Vasantha Rupasinghe *

Department of Plant, Food, and Environmental Sciences, Faculty of Agriculture, Dalhousie University, Truro, NS B2N 5E3, Canada
* Correspondence: vrupasinghe@dal.ca; Tel.: +1-902-893-6623

Abstract: Haskap (*Lonicera caerulea*) berry is rich in anthocyanins, particularly cyanidin-3-*O*-glucoside (C3G). In this investigation, a response surface methodology was applied to optimize the anhydrous ethanol-based extraction parameters to obtain the maximum yield of anthocyanins from haskap berry and to compare the recovery of anthocyanins from different extraction methods. The central composite design was employed to study the effect of three independent variables (X_A = ultrasonic bath power, X_B = extraction temperature, and X_C = extraction time) which were found to significantly affect the response variable total anthocyanin content (TAC) and fit to the second-order polynomial model. The optimum process parameters of X_A = 536 W, X_B = 62.3 °C, and X_C = 63.5 min provided a predicted TAC of 16.5 mg C3G equivalence (C3GE)/g dry weight (DW), which was experimentally validated with 16.1 mg of C3GE/g DW. The optimized ultrasonication-assisted extraction process using anhydrous ethanol was also effective in recovering quercetin glycosides, catechin, procyanidin B2, and iridoids, as determined by ultra-pressure liquid chromatography–mass spectrometry. Though the anthocyanin recovery was the highest (17.6 mg of C3GE/g DW) when a deep eutectic solvent consisting of citric acid and D-(+)-maltose was used, this solvent system has limitations when preparing dehydrated extracts for industrial applications. This study concludes that the effective extraction of anthocyanins and other phytochemicals from haskap berries can be performed using food-grade anhydrous ethanol.

Keywords: blue honeysuckle; polyphenols; cyanidin glucoside; ultrasonication; response surface method; nutraceutical

1. Introduction

Recently, consumer demand has shifted towards health-promoting, natural, and safe foods, reflecting a desire for healthier lifestyles. Therefore, the food industry is exploring innovative supplemented foods and dietary supplements with various bioactive compounds [1]. Plant-food-derived polyphenols have been gaining significant attention among health-promoting food bioactives. In addition, polyphenols have been utilized as natural food additives and food preservatives [2]. Anthocyanins, a sub-group of polyphenols, are abundant in dark-colored berries. Anthocyanin can be used as a food colorant [3] and antioxidant [4]. Therefore, food and nutraceutical industries are interested in identifying natural sources of anthocyanins. Haskap or blue honeysuckle (*Lonicera caerulea*) berry is rich in anthocyanins, particularly cyanidin-3-*O*-glucoside (C3G) [5,6]. The C3G content of haskap berry is significantly higher than that of other berries such as raspberries, blackberries, red currants, and blueberries [6,7]. C3G-rich haskap berry powder and extracts have been reported to possess various potential health benefits including antioxidant, cardioprotective, anti-inflammatory, neuroprotective, anticancer, and anti-diabetic properties [5–10].

Various extraction methods are currently being investigated to extract anthocyanins from plant sources. Anthocyanins are known to be extremely unstable in certain environmental conditions, such as in environments with light, an alkaline pH, oxygen, a high temperature, and the presence of metal ions [11]. Hence, the development of an extraction process that is appropriate for food and nutraceutical industries is crucial. The extraction process should be selected considering the cost, the high recovery of targeted bioactives, the simplicity of evaporation of the solvent, and the safety of solvent residues for human health [12]. The extraction processes under mild or non-thermal processes are required for extracting anthocyanin from haskap berries [13]. These techniques include ultrasound-assisted extraction (UAE), pressurized liquid extraction, microwave-assisted extraction, and enzyme-assisted extraction. These methods are preferred due to their ability to enhance the bioactive yield, reduce the extraction time, operate at low temperatures, minimize the degradation of bioactives during processing, and improve energy efficiency compared to conventional thermal processing [14]. UAE is a cost-effective and efficient method that is used to extract phytochemicals from plant material compared to other methods [15]. A previous study has aimed to optimize various factors, such as solvent-to-solid ratio, solvent composition, ultrasound bath temperature, and the duration of extraction, for the extraction of anthocyanin [7]. The main objective of this study is to develop a scalable, consumer-friendly process to extract anthocyanin from haskap berry to use the extracted anthocyanins in food and nutraceutical applications. Therefore, the primary focus of the present study is to establish the optimal UAE conditions that are required in order to extract anthocyanin from haskap berry using absolute ethanol by varying ultrasonic power along with extraction temperature and duration via a response surface methodology (RSM) using central composite design (CCD). The use of anhydrous ethanol as the extraction solvent is advantageous because it reduces the cost of energy for evaporation to generate dry extract compared to water-based extraction solvents. In addition, ethanol can be recovered when the extracts are dried to generate powder and can be reused for extraction.

2. Results and Discussion

2.1. Model Fitting

The central composite design was used to determine the effect of three extraction parameters (ultrasonic power, extraction temperature, and extraction time) on the total anthocyanin content (TAC) of haskap berries. The levels for each independent variable were selected based on the power range of the ultrasonic bath and the boiling point of the anhydrous ethanol (78 °C). The results of experimental runs are presented in Table 1, while the results of ANOVA are tabulated in Table 2. It is evident that the lack of fit was not significant ($p > 0.05$), indicating that the models could be applied to predict the studied responses. Also, the three-dimensional response surface plots displayed interaction effects between process parameters toward the model responses.

Table 1. Central composite design with coded and actual values for each investigated factor and resultant total anthocyanin content (TAC) (Y in mg C3GE/g DW of haskap berry powder).

Run No.	X_A	X_B	X_C	Y
1	500 (0)	48 (0)	60 (0)	15.8
2	500 (0)	48 (0)	60 (0)	16.3
3	500 (0)	48 (0)	110 (+1.68)	15.9

Table 1. *Cont.*

Run No.	X_A	X_B	X_C	Y
4	500 (0)	48 (0)	10 (−1.68)	12.1
5	260 (−1)	31 (−1)	90 (+1)	12.0
6	740 (1)	65 (1)	90 (+1)	15.2
7	500 (0)	48 (0)	60 (0)	16.5
8	740 (1)	65 (1)	30 (−1)	16.5
9	740 (1)	31 (−1)	30 (−1)	12.9
10	500 (0)	48 (0)	60 (0)	15.3
11	500 (0)	76 (+1.68)	60 (0)	15.6
12	900 (+1.68)	48 (0)	60 (0)	16.1
13	260 (−1)	65 (1)	90 (+1)	16.0
14	100 (−1.68)	48 (0)	60 (0)	13.5
15	260 (−1)	31 (−1)	30 (−1)	11.3
16	260 (−1)	65 (1)	30 (−1)	16.0
17	500 (0)	20 (−1.68)	60 (0)	11.9
18	500 (0)	48 (0)	60 (0)	15.2
19	740 (1)	31 (−1)	90 (+1)	15.5
20	500 (0)	48 (0)	60 (0)	15.9
Predicted optimum conditions	536	62.3	63.5	16.5 *

* Predicted TAC.

Table 2. ANOVA results of significant factors in the quadratic model.

Source	df	Adj SS	Adj MS	F-Value	*p*-Value
Model	9	53.01	5.889	9.10	0.001
Linear	3	35.88	11.96	18.5	0.000
X_A	1	6.074	6.074	9.38	0.012
X_B	1	24.61	24.61	38.0	0.000
X_C	1	5.196	5.196	8.03	0.018
Quadric	3	10.56	3.521	5.44	0.018
X_B. X_B	1	6.369	6.369	9.84	0.011
X_C. X_C	1	4.708	4.708	7.27	0.022
Two-way interaction	3	6.561	2.187	3.38	0.062
X_B. X_A	1	3.736	3.736	5.77	0.037
Error	10	6.472	0.647		
Lack of fit	5	4.997	0.999	3.39	0.103
Pure error	5	1.476	0.295		
Total	19	59.48			

2.2. *Effect of Extraction Variables on TAC*

Based on the obtained responses, after the elimination of all the insignificant effects of extraction variables, the final proposed model for the TAC is shown below (Equation (1)), and the lack of fit was not significant ($p = 0.103$) with the predication of terms ($R^2 = 89.12\%$ and Adj.$R^2 = 79.32\%$), indicating that the model is well fitted and reliable to explore further. Considering the linear effects, ultrasonic power (X_A), extraction temperature (X_B), and extraction time (X_C) had a significant ($p < 0.05$) positive effect on TAC. Based on the quadratic model, a significant ($p < 0.05$) negative effect of extraction temperature and extraction time was observed for the yield of total anthocyanin content (TAC). The interactive effects between extraction temperature and ultrasonic power (X_B. X_A) had a significant ($p < 0.05$) negative effect on the TAC yield.

$$\mathrm{TAC}\left(\frac{\mathrm{mgC3GE}}{\mathrm{gDW}}\right) = -7.09 + 0.01560\ X_A + 0.4689\ X_B + 0.1502\ X_C - 0.002402\ X_B.X_B - 0.000645\ X_C.X_C - 0.000173\ X_B.X_A \quad (1)$$

The response surface plots (Figure 1) show the interactive effects of different variables. The TAC yield for this study varied from 11.30 to 16.54 mg C3GE/g DW. The optimized extraction conditions were X_A = 536 W, X_B = 62.3 °C, and X_C = 63.5 min, with a predicted TAC response of 16.49 mg C3GE/g DW. For the validation of predicted conditions, an experiment was performed using the predicted optimum conditions (X_A = 540 W, X_B = 62 °C, and X_C = 64 min) and the actual TAC of 16.12 mg C3GE/g DW was found. The reported anthocyanin yield when using a deep eutectic solvent (DES) as the extraction solvent and an ultrasound bath temperature of 75 °C for 10 min was 19.6 mg C3GE/g DW [16]. Celli and colleagues [7] reported the total anthocyanin content of 22.73 mg C3GE/g DW when the composition consisted of 80% ethanol, 0.5% formic acid was added to the extraction solvent, and an ultrasound bath temperature of 35 °C was used for 20 min. These reported TAC values are comparatively high compared to the experimental TAC values obtained by this study.

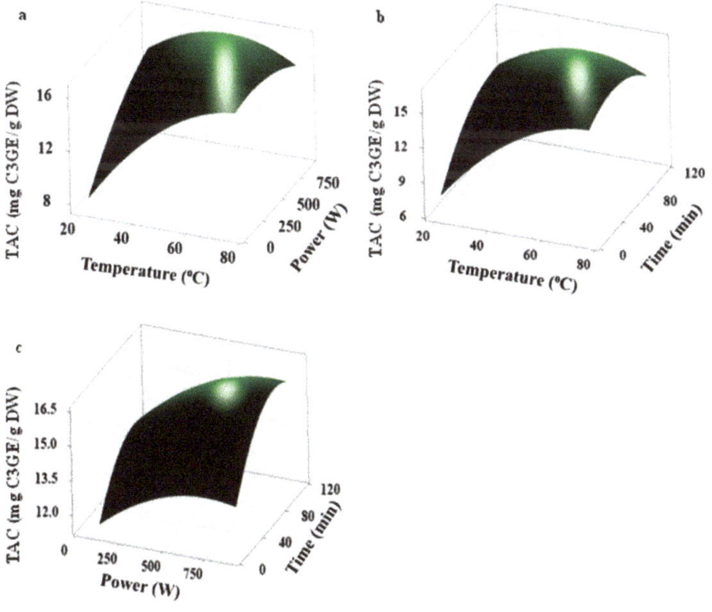

Figure 1. Response surface plots of the TAC (mg C3GE/g DW) extraction from haskap berry powder as a function of ultrasonic power (W), extraction temperature (°C), and extraction time (min). The following values were kept constant: (**a**) extraction time of 60 min, (**b**) ultrasonic power of 500 W, and (**c**) extraction temperature of 48 °C.

Although this study shows high TAC extraction ultrasonic power compared to the highest reported value (510 W, from blueberry) [17], the optimum extraction temperature shows a low value compared to the value reported (75 °C) in another study using the DES as an extraction solvent [16]. Usually, when increasing the ultrasound power until a certain level, higher UAE efficiency can be obtained. However, any further increase in ultrasonication power may cause the degradation of anthocyanins. This degradation can elaborate based on the sonolysis of water stimulated by the cavitation effect during ultrasonication. The cavitation effect could produce hydroxyl radicals (•OH) and hydrogen peroxide (H_2O_2). These reactive oxygen species (ROS) cause the degradation of anthocyanin, forming chalcones by opening the ring structures [18]. When using anhydrous ethanol as an extraction solvent, the generation of ROS may be minimal compared to using other aqueous extraction solvents. Most of the anthocyanin extraction studies conducted had water in their extraction solvents, which may affect the extraction yield at elevated ultrasonication power levels. In the presence of polyphenol oxidase, anthocyanins are enzymatically degraded. This enzyme can be inactivated by heating it up to 50 °C. That way, moderately increased temperature may have a positive effect on anthocyanin retention [19].

2.3. Comparison of the TAC, TPC, Total Antioxidant Capacity (FRAP), DPPH• Radical Scavenging Activity, and Polyphenols of Extracts Prepared from Various Extraction Processes

The low stability of anthocyanins limits their application in the food and nutraceutical industry. The acidification of the extraction solvents may increase the stability of anthocyanin via intermolecular pigmentation [20]. The low pH values, where pH < 6, are beneficial to maintaining the stability of anthocyanins, and the addition of organic acids could improve the anthocyanin stability [21]. The use of DES as an extraction solvent has recently sparked interest as it is a green extraction solvent [14]. Table 3 shows the variation in the TAC and anthocyanin profiles with five different extraction solvents under optimized conditions. In this study, the highest C3G yield was provided by the DES used as the extraction solvent (24.9 mg/g DW), 2% acetic acid or 2% citric acid with anhydrous ethanol, and anhydrous ethanol alone. However, measuring the TAC by pH differential methods shows a significantly higher amount of total anthocyanins in DES extraction compared to the optimized anhydrous ethanol, and deionized water/UAE provided the lowest TAC.

Table 3. Anthocyanin concentrations measured by UPLC-ESI-MS and the pH differential assay of haskap berry extracts prepared using different extraction processes.

Extraction Process	Anthocyanins (mg/g DW) by UPLC-ESI-MS				
	C3G	C35DG	Peo3G	C3R	TAC (mg C3GE/g DW)
Optimum extraction conditions using anhydrous ethanol and UAE (OEE)	23.8 a	0.94 b	1.25 a	1.43 ab	15.9 b
Anhydrous ethanol with 2% Acetic acid/UAE	23.0 a	0.96 b	1.16 b	1.39 b	16.7 ab
Anhydrous ethanol with 2% Citric acid/UAE	22.3 a	0.99 b	1.18 ab	1.43 ab	16.7 ab
DES/UAE	24.9 a	1.13 b	1.16 b	1.57 a	17.6 a
Deionized water/UAE	12.3 b	1.48 a	0.67 c	1.18 c	9.56 c

Mean values of four types of anthocyanin obtained by means of UPLC-ESI-MS analysis (C3G, C35DG, Peo3G, and C3R) (mg/g DW) and the total anthocyanin content measured using the pH differential method (mg C3GE/g DW). Within each column, means sharing the same letters (a–c) are not significantly different ($p < 0.05$). Abbreviations: C3G, cyanidin-3-O-glucoside; C3GE, C3G equivalents, C35DG, cyanidin-3,5-diglucoside; C3R, cyanidin-3-O-rutinoside; DES, deep eutectic solvent system; Peo3G, peonidin-3-O-glucoside.

Among the non-anthocyanin polyphenols investigated, Q3R and chlorogenic acid were detected in relatively high quantities in the five extracts (Table 4). The recovery of Q3R was greater in all anhydrous ethanol extraction processes compared to water extractions. The addition of 2% acetate or 2% citrate to absolute ethanol did not significantly increase the TPC of the extracts (Figure 2A).

Table 4. Flavonoid and chlorogenic acid content of haskap berry extracts prepared from different extraction processes.

Extraction Process	Flavonoids Concentration by UPLC-ESI-MS (mg/g DW)						
	Q3G	Q3ArG	Quercetin	Q3R	Catechins	PB2	ChlA.
Optimum extraction conditions using anhydrous ethanol/UAE (OEE)	0.33 a	0.13 b	0.01 ab	1.73 b	2.14 b	0.10 a	1.69 a
Anhydrous ethanol with 2% acetic acid/UAE	0.26 b	0.11 b	0.01 ab	1.44 c	2.37 b	0.09 a	1.76 a
Anhydrous ethanol with 2% citric acid/UAE	0.35 a	0.15 a	0.02 a	2.00 a	2.82 a	0.11 a	1.84 a
DES/UAE	0.22 b	0.13 b	0.01 bc	1.41 c	3.17 a	0.09 a	1.93 a
Deionized water/UAE	0.13 c	0.11 b	0.00 c	1.22 c	0.98 c	0.08 a	1.33 b

Mean values of different flavonoids quantified by means of UPLC-ESI-MS analysis (mg/g DW). Within each column, means sharing the same letters (a–c) are not significantly different ($p < 0.05$). Catechins are composed of epicatechin and catechin. Abbreviations: DES, deep eutectic solvent system; Q3G, quercetin-3-*O*-galactoside; Q3ArG, quercetin-3-arabinoglucoside (peltatoside); Q3R, quercetin-3-*O*-rutinoside; PB2, proanthocyanin B2; ChlA., chlorogenic acid; and TPC, total phenolic content.

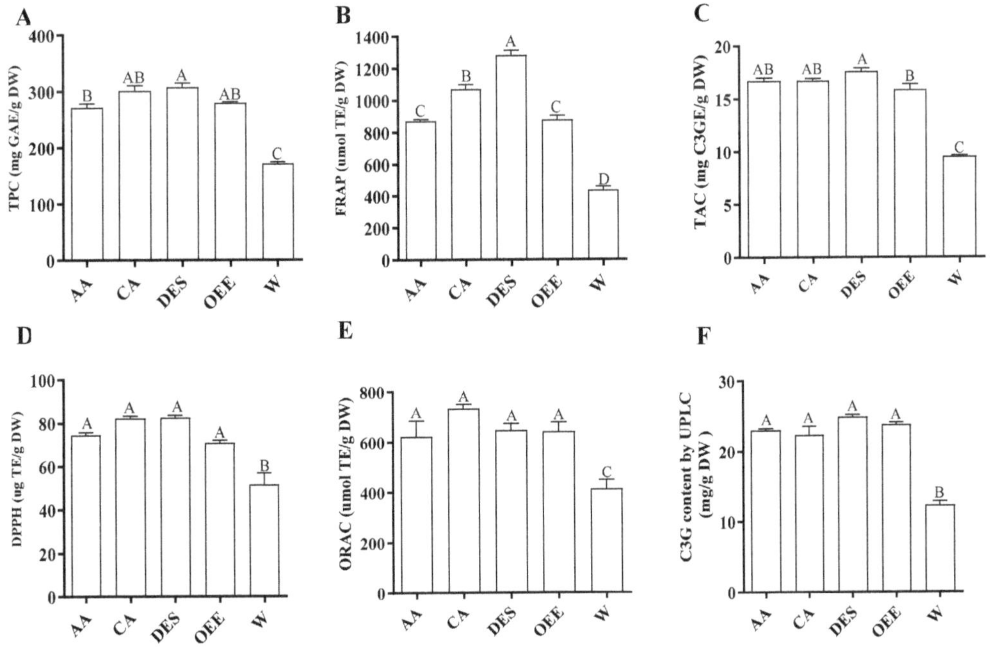

Figure 2. The TPC (**A**), FRAP (**B**), TAC (**C**), DPPH (**D**), ORAC (**E**), and C3G (**F**) content variations of different extraction processes. Abbreviations: C3GEs, C3G equivalents, OEE, optimum extraction condition using anhydrous ethanol and UAE; AA, anhydrous ethanol with 2% acetic acid/UAE); CA, anhydrous ethanol with 2% citric acid/UAE; DES, deep eutectic solvent system with UAE; W, deionized water with UAE. The different letters of means (A–D) indicate the significant differences ($p > 0.05$) among them.

The DPPH radical scavenging activity of different extraction methods was investigated based on the inhibition percentage (Figure 2D). When water was used as the extraction solvent, it showed the least antioxidant activity and when the DES was used as the extraction solvent, the highest antioxidant activity of the anthocyanin was observed. However, the DPPH radical scavenging activity of the DES was not significantly different from that of extracts prepared using 2% acetic acid in ethanol and 2% citric acid in ethanol. When water was used as the extraction solvent, it showed the least TAC values (Figure 2C). However, the ORAC value of optimized anhydrous ethanol extract was not significantly different from that of extracts prepared using 2% acetic acid in ethanol and 2% citric acid in ethanol and the DES.

C3G comprises about 80% anthocyanins of haskap beery [5]. C3G is found in almost all North American berries. C3G content is significantly greater in haskap berries compared to other common berries such as strawberries, blueberries, cranberries, and chokeberries [6]. The total phenolic content of haskap berry was reported to be the highest (1111 mg GAE/100 g FW) among all other North American berry types [22].

Anhydrous ethanol was effective in recovering polyphenols with a relatively higher antioxidant capacity than water (Figure 2). In general, aqueous ethanol has been commonly used for the extraction of polyphenols from plant materials [23]. Celli and colleagues [7] extracted C3G, C3R, and Peo3G using 80% aqueous ethanol with 0.5% formic acid as the extraction solvent. Since the objective of this study was to develop a scalable, consumer-friendly process to extract anthocyanin from haskap berries, it is mandatory to use a solvent or a solvent mixture that is not toxic to human health. Residues of toxic organic solvents are a major concern in food and nutraceutical industries. The ingestion of formic acid in the digestive tract of humans can cause ulcers and formic acid to metabolize to formaldehyde [24]. As a solvent that is generally recognized as safe (GRAS), anhydrous ethanol [25] shows relatively low toxicity toward human health when compared to organic solvents such as methanol, acetone, hexane, formic acid, and acetonitrile [26,27]. For the commercial production of dried food ingredients or nutraceuticals, the evaporation of extracts is required [28]. Due to the low boiling point of anhydrous ethanol, it can be easily evaporated with a low energy cost compared to water or aqueous ethanol-based extracts [29]. The maltose of DES-based extracts in supplemented food could be a negative factor for consumers with type 2 diabetes [30].

2.4. Correlation Analysis of the TAC, TPC, C3G, ORAC, DPPH, and FRAP of Haskap Berry Extracts

To understand the correlations among the measured parameters, the Pearson correlation analysis of the TAC, TPC, C3G, ORAC, DPPH, and FRAP was performed (Figure 3). The strongest correlation can be seen between the TAC and TPC, which agrees with a previous study [31]. A strong correlation between the TAC and C3G was also observed, demonstrating the contribution of C3G to total anthocyanins. A similar correlation has been shown by Khattab et al. [32]. Strong correlations were shown between the TPC and three antioxidant capacity assays (ORAC, FRAP, and DPPH). DPPH results showed a strong correlation with the FRAP, TAC, and TPC. Antioxidant capacity from the FRAP showed a strong correlation ($R = 0.93$) with the TPC, but the ORAC correlation with the FRAP was less strong ($R = 0.73$) compared to the correlation between other antioxidant assays. Differences in antioxidant capacities in different assays may be attributed to the variations in the mechanisms underlying the assays [33].

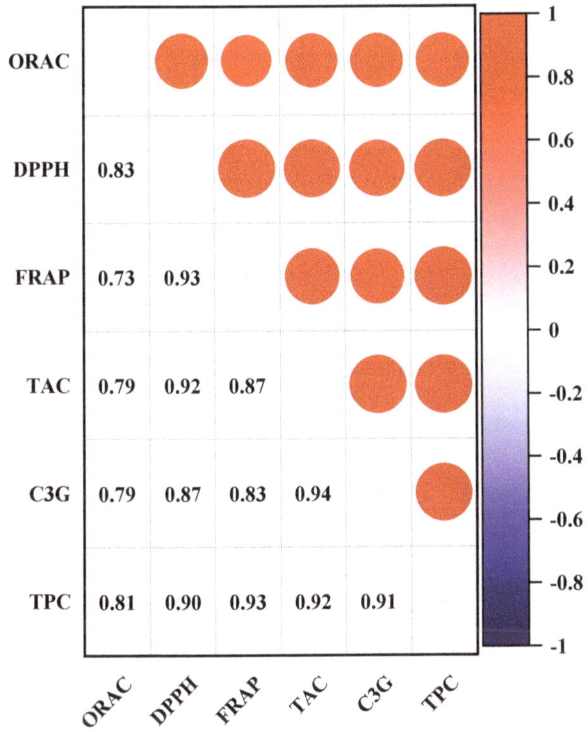

Figure 3. Correlation coefficients among the TAC, TPC, C3G, ORAC, DPPH, and FRAP of haskap berry extracts prepared from various methods (n = 15).

2.5. Iridoid Content of Haskap Berry Extracts

Iridoids are potent with many biological activities, such as anti-inflammatory, antitumor, hepatoprotective, neuroprotective, and cardioprotective properties [34]. Structurally, iridoids can be divided into four groups: iridoid glycosides, secoiridoid glycosides, non-glycosidic iridoids, and bis-iridoids. Iridoids investigated in this study mainly belong to iridoid glycosides (loganin and loganic acid), and secoiridoid glycoside (sweroside) [34].

The highest total iridoids (loganin, loganic acid, and sweroside) were achieved using deionized water and a DES as the extraction solvents (Table 5). The recovery of loganin was not influenced by the different extraction methods used in this study. In contrast, sweroside extraction was most effective using the DES, which may be due to increased solvent polarity by citric acid. Similar to our findings, loganic acid was the most abundant iridoids in haskap berries, representing 22% to 73% of the total iridoids [35]. Hot water extraction recovered more iridoids than ethanol or methanol [36]. While polyphenols are prevalent in fruits, iridoids are relatively rare, exceptions include cranberry [37], bilberry [38], and haskap berries [35]. Cornelian cherry contains a wide range of iridoid content from 87 to 493 mg/100 g FW [39]. In 13 wild blueberry genotypes, iridoids range from 2.0 to 237 mg/100 g DW [40]. Bilberry contains 13 mg/100 g FW [41]. The previously reported iridoid contents of haskap berry grown in Poland are much higher (120 to 276 mg/100 g FW) [35] compared to the Canadian-grown haskap berries used in this study.

Table 5. Iridoid content of haskap berry extracts.

Extraction Process	Iridoids Concentration by UPLC-ESI-MS (mg/g DW)			
	Loganin	Loganic Acid	Sweroside	Total Iridoids
Optimum extraction conditions using anhydrous ethanol/UAE (OEE)	0.23 a	1.27 b	0.23 d	1.72 b
Anhydrous ethanol with 2% acetic acid/UAE	0.21 a	1.22 b	0.23 cd	1.67 b
Anhydrous ethanol with 2% citric acid/UAE	0.22 a	0.60 c	0.26 bc	1.08 c
DES/UAE	0.22 a	1.93 a	0.31 a	2.46 a
Deionized water/UAE	0.20 a	1.94 a	0.26 bc	2.41 a

Mean values of most abundant iridoids (loganin, loganic acid, and sweroside) measured by UPLC-ESI-MS analysis (mg/g DW). Within each column, the means sharing the same letters (a–d) are not significantly different ($p < 0.05$). Total iridoids are composed of loganin, loganic acid, and sweroside. Abbreviations: DES, deep eutectic solvent system.

2.6. Single-Factor Experiment—The Effect of Ethanol and Ultrasonication Power on Anthocyanin Extraction

When water was used as an extraction solvent, the lowest amount of anthocyanins was recovered (Figure 4A). Ethanol percentage ranging from 40% to 90% recovered the highest amount of anthocyanins. A similar pattern of recovery of anthocyanins was observed by Cacace et al. [42] when extracting anthocyanin from black currant berries. The electric charge changes with the addition of ethanol are responsible for this behavior. The extraction yields of polyphenols were affected by the concentration of ethanol in the solvent. When the ethanol concentration is high, the dielectric constant is low, so it lowers the energy required to break the water arrangement [43]. Therefore, nonpolar flavanols needed a higher ethanol concentration (75% to 100%) to obtain a high recovery. For polar phenolic acids, the optimum concentration of ethanol was 70% to 90%. On the other hand, a mixture of ionic anthocyanins (flavylium cation, AH^+) did not require a high ethanol concentration, and the maximum yield can be obtained at approximately 50% ethanol [42].

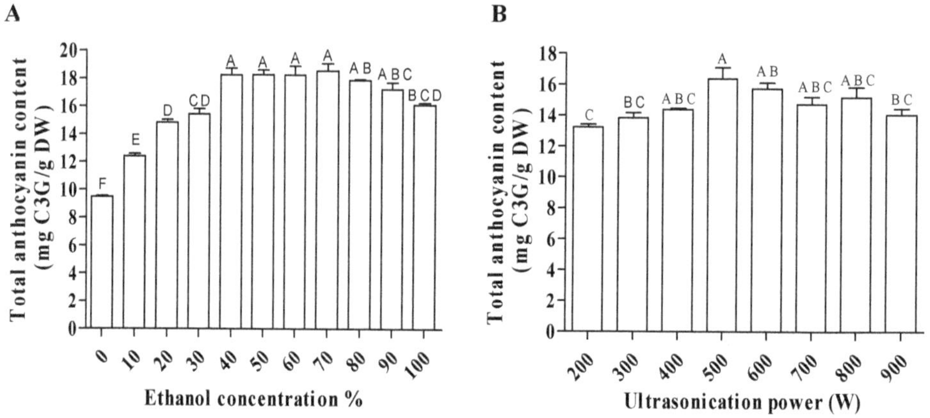

Figure 4. TAC of haskap berry extracts in response to different ethanol concentrations (**A**) and different ultrasonic power levels (**B**). The different letters of means (A–D) indicate the significant differences ($p > 0.05$) among them.

The highest TAC can be obtained using the ultrasonication power ranging from 400 W to 800 W (Figure 4B). Very few studies have investigated the impact of ultrasonication power on anthocyanin recovery. One study showed that 400 W gives the maximum anthocyanin recovery from *Rubia sylvatica* [44]. Liao et al. [45] described 200 W as the power which contributes to the highest anthocyanin yield from purple eggplant peel. This study was carried out using a power range of 100 W to 300 W. Our study shows that increasing the power above 400 W is not necessary in terms of energy cost for the process, as the anthocyanin yield is not significantly different when ultrasonic power of 400 W is used.

2.7. TAC Recovery Percentage at Optimum Extraction Conditions

The recovery percentage of the TAC from haskap powder was performed in subsequent different extraction attempts using the same optimum conditions until anthocyanin color was not visually observed in the supernatant (Table S1). Based on the TAC determinations, the first attempt recovered about 89.0% of the potential total TAC. Interestingly, 97.9% of the TAC can be recovered by conducting two subsequent extractions and pooling them. The above observation indicates that the optimized extraction process recovers about 90% of extractable anthocyanins from dehydrated haskap berry powder.

3. Material and Methods

3.1. Plant Materials, Chemicals, Reagents, and Equipment

The fully mature haskap berry variety 'Tundra' of 2022 harvest was obtained from Atlantic Haskap, Elderbank, NS, Canada. Gallic acid, sodium carbonate, sodium acetate trihydrate, 2,4,6-tris(2-pyridyl)-*S*-triazine (TPTZ), Folin–Ciocalteu reagent, ferric chloride, 6-hydroxy-2,5,7,8-tetramethylchroman-2-carboxylic acid (Trolox), D-(+)-maltose, citric acid, glacial acetic acid, and 1,1-diphenyl-2-picrylhydrazyl (DPPH) were obtained from Sigma-Aldrich (Oakville, ON, Canada). Anhydrous ethanol was purchased from Commercial Alcohols, Brampton, ON, Canada. The extraction process was conducted in an ultrasonic bath with a fixed frequency of 40 kHz (Model 4HT-1524-12, Crest, Ewing, NJ, USA). The temperature was controlled within ±1 °C with a calibrated thermometer and adjusted with cold water.

3.2. Experimental Design

The experiment was conducted in two stages. In the first stage, an RSM was selected for the optimization of process parameters to maximize the yield of total anthocyanin content (TAC) from freeze-dried haskap berry powder (particle size 0.1 to 10 μm), which was prepared using a freeze dryer (Kinetics, FTS Systems Inc., Stone Ridge, NY, USA) at −40 °C for 72 h and then ground using a standard coffee grinder. A central composite design was specifically chosen to study the effect of three independent variables (X_A = ultrasonic bath power (W), X_B = extraction temperature (°C), and X_C = extraction time (min)) on the response variables (Y) of the TAC (mg C3GE/g DW). The ranges of independent variables used for the optimization study were as follows: X_A—100 W to 900 W, X_B—20 °C to 76 °C, and X_C—10 min to 110 min. This study explored the effect of these variables on anthocyanin extraction using a solvent-to-solid ratio of 25:1 mg/mL [7] and a solvent composition consisting of anhydrous ethanol. A level-three factor central composite design, consisting of 20 experimental runs, was performed, and the observed values are given in Table 1. Response surface analysis and analysis of variance (ANOVA) were used to determine the regression coefficients and the statistical significance of the model terms.

The observed values were then fitted into a second-order polynomial model, as shown in Equation (2), and the regression coefficient β was generated, as shown in Table 2.

$$Y = \beta_0 + \Sigma\beta_i X_i + \Sigma\beta_{ii} X_i 2 + \Sigma\beta_{ij} X_{ij} \quad (2)$$

Y indicates the response variable; X_i and X_j are independent variables; β_0 is the constant coefficient of the model; and β_i, β_{ii}, and β_{ij} are the regression coefficients of the single effects, quadratic effects, and interactive effects of independent variables, respectively.

In the second phase used for drawing comparisons with anhydrous ethanol (AE), four different extraction solvents, 2% acetic acid in anhydrous ethanol (AA), 2% citric acid in anhydrous ethanol (CA), water (W), and deep eutectic solvent (DES) consisting of citric acid and D-(+)-maltose [16], were used with the same optimal parameters of this study. The second-phase extracts were analyzed to identify the TAC, the total phenolic content (TPC), and total antioxidant capacity using the ferric-reducing antioxidant power (FRAP) assay, the DPPH radical scavenging assay, the oxygen radical absorbance capacity assay (ORAC), and polyphenol profiling using ultra-pressure liquid chromatography (UPLC) with electrospray ionization (ESI) and mass spectrometry (MS) [5].

3.3. *Validation of the Model*

The optimized conditions were validated by measuring the TAC of extracts prepared based on the optimum conditions predicted using an RSM in Minitab® (v. 20). The experimental values were compared with predicted values to determine the validity of the model.

3.4. *UAE of Anthocyanins*

UAE was conducted in 50 mL amber glass vials, where 0.5 g of freeze-dried haskap berry powder was transferred, and 12.5 mL of extraction solvent was added to each vial. The vials were vortexed for 10 s and placed in the ultrasonic bath for each extraction condition. Following extraction, samples were centrifuged (Sorvall ST 16, Thermo Scientific, Darmstadt, Germany) at 3000× *g* for 5 min. The supernatant was then filtered through a 0.45 µm syringe filter and stored in 15 mL Falcon tubes in the dark at −20 °C.

3.5. *Determination of Total Monomeric Anthocyanin Content (TAC)*

The TAC of samples was determined in triplicate using the pH differential method (AOAC method 2005.02) using a UV–visible spectrophotometer (Tecan Infinite® M200 PRO, Morrisville, NC, USA). Thirty-times-diluted samples in both pH 1 and pH 4.5 buffer were prepared, and the absorbance (A) was measured at 520 nm and 700 nm.

The A of the extract was calculated using the following equation:

$$A = (A520\text{–}A700) \text{ at pH } 1.0 - (A520\text{–}A700) \text{ at pH} \quad (3)$$

The TAC was calculated using the following equation:

$$TAC = (A \times M \times DF \times 103)/(\varepsilon \times L) \quad (4)$$

This equation shows the molar extinction coefficient (ε), 26,900; the molecular weight of C3G (MW), 484.83 g/mol; the dilution factor (DF); and the path length (L) [46]. Results are expressed in mg of C3G equivalents per g of the dry weight (mg C3GE/g DW) of haskap berries.

3.6. Determination of Polyphenols of the Optimized Extract Using UPLC-ESI-MS Analysis

The most prominent polyphenols present in the haskap extracts were identified and quantified [5] using UPLC-ESI-MS (Quattro Micro, Waters, Milford, MA, USA). Briefly, extracts were filtered through 0.45 μm nylon syringe filters. For the separation of analytes, an Aquity BEH C_{18} (100 mm × 2.1 mm, 1.7 μm) column (Waters, Milford, MA, USA) was used with an ESI in positive ion mode (ESI+), with a capillary voltage of 3000 V and a nebulizer gas (N_2) temperature of 375 °C. A flow rate of 0.2 mL/min using 2% formic acid in water (A) and 2% formic acid in methanol was used. A linear gradient was used with the following proportions of solvent B (time in min [t], B%): (0, 10%), (7, 30%), (16, 40%), (17, 40%), (18, 90%), and (20, 10%). The anthocyanins ($M + H^+$) were identified using single-ion monitoring (SIM) mode: m/z 448.8 for cyanidin-3-glucoside, m/z 610.8 for cyanidin-3,5-diglucoside, m/z 462.8 for peonidin-3-glucoside, and m/z 594.8 for cyanidin-3-rutinoside. The analyses of non-anthocyanins (other polyphenols and iridoids) were performed using ESI in negative ion mode (ESI−), with a capillary voltage of 3000 V and a nebulizer gas (N_2) temperature of 375 °C. The same column was used with a flow rate of 0.3 mL/min and a gradient mobile phase using 0.1% formic acid in water (A) and 0.1% formic acid in acetonitrile (B). Solvent B was applied at time t (min) for the following (t, B%): (0, 6%), (6, 20%), (8, 80%), (9, 80%), and (10, 6%). The following analytes were identified using SIM mode: m/z 300.7 for quercetin (Q), m/z 462.5 for Q-3-glucoside, m/z 608.5 for Q-3-rutinoside, m/z 594.7 for Q-3-arabinoglucoside (peltatoside), m/z 288.7 for catechin, m/z 577.1 for procyanidin B2, m/z 352.7 for chlorogenic acid, m/z 435.1 for loganin, m/z 375.1 for loganic acid, and m/z 403.1 for sweroside. All the analytes were quantified using calibration curves generated using external standards. The limit of detection of all the analytes was between 0.01 and 0.1 mg/L.

3.7. Determination of Total Phenolic Content (TPC)

The TPC in extracts was measured using the Folin–Ciocalteu (FC) assay, as described by Huber et al. [47]. Gallic acid was used for the generation of a standard curve using the extraction solvent (anhydrous ethanol) and diluted from concentrations of 59 to 1470 μM. The solutions were made fresh under reduced light conditions and the reaction was carried out under dark conditions. Then, 20 mL of the diluted extract or gallic acid standard was mixed with 100 μL of the 0.2 N Folin–Ciocalteu phenol reagent in 96-well, clear, polystyrene microplates (COSTAR 9017) and gently mixed. After 5 min, 80 μL of 7.5% (*w*/*v*) sodium carbonate was added to each well and mixed. The mixture was incubated for 2 h at ambient temperature before the absorption was measured at 760 nm using the Tecan Infinite® M200 PRO (Morrisville, NC, USA). Results were expressed as mg of gallic acid equivalents per g of dry weight (mg GAE/g DW).

3.8. The Ferric-Reducing Antioxidant Power (FRAP) Assay

The FRAP assay was performed according to Benzie et al. [48] with some modifications. The reaction reagent (FRAP solution) was made immediately before the assay by mixing 300 mM acetate buffer (pH 3.6), 10 mM TPTZ solution, and 20 mM ferric chloride solution in a ratio of 10:1:1. The Trolox standard solution was prepared by dissolving 0.025 g of Trolox in 100 mL of anhydrous ethanol to make 1 mM Trolox. A calibration curve was developed using Trolox stock solution with appropriately anhydrous ethanol dilutions (from 5 to 450 μM Trolox concentrations). Both the FRAP solution and the samples in the microplate were warmed to 37 °C prior to the assay. The FRAP analysis was performed by reacting 20 μL of blank or standard sample with a 180 μL FRAP solution in 96-well clear polystyrene plates (COSTAR 9017). Absorption was measured at 593 nm using the Tecan

Infinite® M200 PRO (Morrisville, NC, USA). FRAP values were expressed as μmol Trolox equivalents per g of the sample dry weight (μmol TE/g DW).

3.9. DPPH Radical Scavenging Assay

The free radical scavenging activity of different anthocyanin extracts was measured using the method described by Guo et al. [49] with slight modifications. In short, the DPPH solution was made by dissolving 7.9 mg of DPPH in 100 mL of methanol and stored at −20 °C. The DPPH solution (100 μL) and the sample (100 μL) were mixed in a 96-well plate and kept in a 37 °C incubator for 30 min. The absorbance was measured at 520 nm using the Tecan Infinite® M200 PRO (Morrisville, NC, USA). The Trolox standard stock solution was prepared by dissolving 0.25 mg of Trolox in 100 mL of the anhydrous ethanol to make 2.5 mg/L Trolox. A calibration curve was developed using the Trolox stock solution with appropriate dilutions (from 0.781 to 25 μg/L Trolox concentrations). The following equation was used to calculate the inhibition percentage:

$$\text{Antioxidant activity (\%inhibition)} = \left(\frac{\text{Ab blank} - \text{Ab sample}}{\text{Ab blank}}\right) \times 100 \quad (5)$$

where Ab sample is the absorbance value of different anthocyanin extracts and Ab blank is the absorbance value of the anthocyanin-free extract solvent. The antioxidant activity was expressed as μg Trolox equivalents per g of the sample dry weight (μg TE/g DW).

3.10. Oxygen Radical Absorbance Capacity (ORAC) Assay

The ORAC assay was performed using a previously established protocol [50]. The samples and standards of the fluorescein sodium salt (0.957 mM) were dissolved in 75 mM phosphate buffer (K_2HPO_4/NaH_2PO_4, pH 7). From the sample or Trolox standard, 35 μL and 130 μL volumes of the fluorescein probe were combined in black 96-well polystyrene microplate wells and the plate was warmed to 37 °C for 10 min. After incubation, 35 μL of 150 mM pre-warmed (37 °C) AAPH was transferred into the wells. The plate was maintained at 37 °C for the duration of the analyses (approximately 200 min) with excitation and emission readings every 60 s. Excitation of the reaction mixture was carried out at 490 nm and the emission was read at 540 nm. The antioxidant capacity of the samples was calculated as Trolox equivalents using a quadratic relation developed from the area under the fluorescence decay curves for standards made from 10 to 150 μM concentrations. The antioxidant activity was expressed as μmol Trolox equivalents per g of the sample dry weight (μmole TE/g DW).

3.11. Single-Factor Experiment

The effect of ethanol percentage and ultrasonication power on the TAC of extracts was studied using single-factor experiments. One factor was changed while the other factors were fixed in each experiment at optimum extraction conditions (540 W, 62 °C, and 64 min). The effect of ultrasound power was studied at 200–900 W while the other factors (a temperature of 62 °C and a time of 64 min) remained fixed. In the same manner, the ethanol concentration from 0% to 100% was investigated at 540 W, 62 °C, and 64 min. The extracts were centrifuged, and the supernatants were tested for the TAC.

4. Conclusions

This study investigated the optimized conditions required for the recovery of anthocyanins from freeze-dried haskap berry powder using anhydrous ethanol as an extraction solvent. Optimized conditions for UAE for total anthocyanin extraction were a temperature of 62.3 °C, an extraction time of 63.5 min, and an ultrasonic power of 536 W. The validity of the model obtained by the RSM was demonstrated by the similar extraction yield to the predicted response (16.5 mg C3GE/g DW) with the actual experimental result (16.1 mg C3GE/g DW) under optimum extraction conditions. The addition of citric acid or acetic acid to the anhydrous ethanol does not affect the extraction yield of C3G from haskap berry powder under the optimized conditions. The emerging DES containing maltose and citric acid as an alternative extraction process is as effective as optimized anhydrous ethanol extraction; however, the DES has limitations in scale-up extractions for commercial applications. Compared to traditional solvents such as methanol, food-grade ethanol has a low environmental impact due to fast biodegradation and is less harmful to the environment. The findings of the current study are useful as they encourage the food and nutraceutical industry to use food-grade anhydrous ethanol as a cost-effective and green extraction solvent to extract and prepare C3G-rich bioactive preparations from haskap berries.

Supplementary Materials: The following supporting information can be downloaded at the following: https://www.mdpi.com/article/10.3390/molecules30051071/s1, Tabel S1. Anthocyanin recovery percentage at optimum extraction condition.

Author Contributions: Conceptualization, D.C. and H.P.V.R.; Methodology, D.C. and H.P.V.R.; Validation, D.C.; Formal analysis, D.C.; Investigation, D.C. and H.P.V.R.; Resources, H.P.V.R.; Data curation, H.P.V.R.; Writing—original draft, D.C.; Writing—review & editing, H.P.V.R.; Supervision, H.P.V.R.; Project administration, H.P.V.R.; Funding acquisition, H.P.V.R. All authors have read and agreed to the published version of the manuscript.

Funding: We thank the Schwabe Phyto Innovation Challenge program of Dr. Willmar Schwabe GmbH & Co KG, Wilmar-Schwabe-Str., Karlsruhe, Germany, for providing financial support.

Institutional Review Board Statement: Not applicable.

Informed Consent Statement: Not applicable.

Data Availability Statement: Data are contained within the article.

Acknowledgments: We thank Atlantic Haskap, Elderbank, NS, Canada for providing the haskap berries for this study.

Conflicts of Interest: The authors declare no conflict of interest.

References

1. Kussmann, M.; Abe Cunha, D.H.; Berciano, S. Bioactive Compounds for Human and Planetary Health. *Front. Nutr.* **2023**, *10*, 1193848. [CrossRef] [PubMed]
2. Pop, O.L.; Suharoschi, R.; Socaci, S.A.; Berger Ceresino, E.; Weber, A.; Gruber-Traub, C.; Vodnar, D.C.; Fărcaș, A.C.; Johansson, E. Polyphenols—Ensured Accessibility from Food to the Human Metabolism by Chemical and Biotechnological Treatments. *Antioxidants* **2023**, *12*, 865. [CrossRef] [PubMed]
3. Bendokas, V.; Stanys, V.; Mažeikienė, I.; Trumbeckaite, S.; Baniene, R.; Liobikas, J. Anthocyanins: From the Field to the Antioxidants in the Body. *Antioxidants* **2020**, *9*, 819. [CrossRef] [PubMed]
4. Sadowska-Bartosz, I.; Bartosz, G. Antioxidant Activity of Anthocyanins and Anthocyanidins: A Critical Review. *Int. J. Mol. Sci.* **2024**, *25*, 12001. [CrossRef] [PubMed]
5. De Silva, A.B.K.H.; Rupasinghe, H.P.V. Polyphenols Composition and Anti-Diabetic Properties in Vitro of Haskap (*Lonicera caerulea* L.) Berries in Relation to Cultivar and Harvesting Date. *J. Food Compos. Anal.* **2020**, *88*, 103402. [CrossRef]
6. Rupasinghe, H.P.V.; Arumuggam, N.; Amararathna, M.; De Silva, A.B.K.H. The Potential Health Benefits of Haskap (*Lonicera caerulea* L.): Role of Cyanidin-3-O-Glucoside. *J. Funct. Foods* **2018**, *44*, 24–39. [CrossRef]
7. Celli, G.B.; Ghanem, A.; Brooks, M.S.-L. Optimization of Ultrasound-Assisted Extraction of Anthocyanins from Haskap Berries (*Lonicera caerulea* L.) Using Response Surface Methodology. *Ultrason. Sonochem.* **2015**, *27*, 449–455. [CrossRef] [PubMed]
8. Amararathna, M.; Hoskin, D.W.; Rupasinghe, H.P.V. Cyanidin-3-O-Glucoside-Rich Haskap Berry Administration Suppresses Carcinogen-Induced Lung Tumorigenesis in A/JCr Mice. *Molecules* **2020**, *25*, 3823. [CrossRef]
9. Amararathna, M.; Hoskin, D.W.; Goralski, K.B.; Rupasinghe, H.P.V. Suppression of NNK Metabolism by Anthocyanin-Rich Haskap Berry Supplementation Through Modulation of P450 Enzymes. *Pharmaceuticals* **2024**, *17*, 1615. [CrossRef] [PubMed]

10. Biswas, D.; De Silva, A.B.K.H.; Mercer, A.; Sarkar, S.; Kienesberger, P.; Langille, M.; Rupasinghe, H.P.V.; Pulinilkunnil, T. Supplementation of Cyanidin-3-*O*-β-Glucoside-Rich Haskap (*Lonicera caerulea* L.) Berry Extract Attenuates Hepatic Lipid Dysregulation in Diet-Induced Obese Mice. *J. Funct. Foods* **2023**, *108*, 105635. [CrossRef]
11. Pace, E.; Jiang, Y.; Clemens, A.; Crossman, T.; Rupasinghe, H.P.V. Impact of Thermal Degradation of Cyanidin-3-*O*-Glucoside of Haskap Berry on Cytotoxicity of Hepatocellular Carcinoma HepG2 and Breast Cancer MDA-MB-231 Cells. *Antioxidants* **2018**, *7*, 24. [CrossRef]
12. Chemat, F.; Khan, M.K. Applications of Ultrasound in Food Technology: Processing, Preservation and Extraction. *Ultrason. Sonochem.* **2011**, *18*, 813–835. [CrossRef] [PubMed]
13. Tan, J.; Han, Y.; Han, B.; Qi, X.; Cai, X.; Ge, S.; Xue, H. Extraction and Purification of Anthocyanins: A Review. *J. Agric. Food Res.* **2022**, *8*, 100306. [CrossRef]
14. Valisakkagari, H.; Chaturvedi, C.; Rupasinghe, H.P.V. Green Extraction of Phytochemicals from Fresh Vegetable Waste and Their Potential Application as Cosmeceuticals for Skin Health. *Processes* **2024**, *12*, 742. [CrossRef]
15. Saifullah, M.; McCullum, R.; McCluskey, A.; Vuong, Q. Comparison of Conventional Extraction Technique with Ultrasound Assisted Extraction on Recovery of Phenolic Compounds from Lemon Scented Tea Tree (*Leptospermum petersonii*) Leaves. *Heliyon* **2020**, *6*, e03666. [CrossRef] [PubMed]
16. MacLean, A.M.G.; Silva, Y.P.A.; Jiao, G.; Brooks, M.S. Ultrasound-Assisted Extraction of Anthocyanins from Haskap (*Lonicera caerulea* L.) Berries Using a Deep Eutectic Solvent (DES) DES Extraction of Anthocyanins from Haskap Berries. *Food Technol. Biotechnol.* **2021**, *59*, 56–62. [CrossRef]
17. Tian, Y.; Yang, Y.; Gao, P.; Wang, J.H.; Xu, Y.Q.; Yu, Z.Y. Optimization of Ultrasonic-Assisted Extraction of Flavonols and Anthocyanins from Blueberry Using RSM. *Adv. Mater. Res.* **2012**, *468*, 2423–2430. [CrossRef]
18. Tiwari, B.K.; Patras, A.; Brunton, N.; Cullen, P.J.; O'Donnell, C.P. Effect of Ultrasound Processing on Anthocyanins and Color of Red Grape Juice. *Ultrason. Sonochem.* **2010**, *17*, 598–604. [CrossRef] [PubMed]
19. Patras, A.; Brunton, N.P.; O'Donnell, C.; Tiwari, B.K. Effect of Thermal Processing on Anthocyanin Stability in Foods; Mechanisms and Kinetics of Degradation. *Trends Food Sci. Technol.* **2010**, *21*, 3–11. [CrossRef]
20. Nistor, M.; Diaconeasa, Z.; Frond, A.D.; Stirbu, I.; Socaciu, C.; Pintea, A.; Rugina, D. Comparative Efficiency of Different Solvents for the Anthocyanins Extraction from Chokeberries and Black Carrots, to Preserve Their Antioxidant Activity. *Chem. Pap.* **2021**, *75*, 813–822. [CrossRef]
21. Liu, Y.; Liu, Y.; Tao, C.; Liu, M.; Pan, Y.; Lv, Z. Effect of Temperature and pH on Stability of Anthocyanin Obtained from Blueberry. *J. Food Meas. Charact.* **2018**, *12*, 1744–1753. [CrossRef]
22. Bakowska-Barczak, A.M.; Marianchuk, M.; Kolodziejczyk, P. Survey of Bioactive Components in Western Canadian berriesThis Article Is One of a Selection of Papers Published in This Special Issue (Part 2 of 2) on the Safety and Efficacy of Natural Health Products. *Can. J. Physiol. Pharmacol.* **2007**, *85*, 1139–1152. [CrossRef] [PubMed]
23. Vieito, C.; Fernandes, E.; Velho, M.V.; Pires, P. The Effect of Different Solvents on Extraction Yield, Total Phenolic Content and Antioxidant Activity of Extracts from Pine Bark (*Pinus pinaster* Subsp. Atlantica). *Chem. Eng. Trans.* **2018**, *64*, 127–132. [CrossRef]
24. Garai, P.; Banerjee, P.; Sharma, P.; Chatterjee, A.; Bhattacharya, R.; Saha, N.C. Mechanistic Insights to Lactic and Formic Acid Toxicity on Benthic Oligochaete Worm Tubifex Tubifex. *Environ. Sci. Pollut. Res.* **2022**, *29*, 87319–87333. [CrossRef] [PubMed]
25. Pimentel-Moral, S.; Borrás-Linares, I.; Lozano-Sánchez, J.; Alañón, M.E.; Arráez-Román, D.; Segura-Carretero, A. Pressurized GRAS Solvents for the Green Extraction of Phenolic Compounds from *Hibiscus sabdariffa* Calyces. *Food Res. Int.* **2020**, *137*, 109466. [CrossRef] [PubMed]
26. Joshi, D.R.; Adhikari, N. An Overview on Common Organic Solvents and Their Toxicity. *J. Pharm. Res. Int.* **2019**, *28*, 1–18. [CrossRef]
27. Liesivuori, J.; Savolainen, H. Methanol and Formic Acid Toxicity: Biochemical Mechanisms. *Pharmacol. Toxicol.* **1991**, *69*, 157–163. [CrossRef] [PubMed]
28. Lapkin, A.A.; Plucinski, P.K.; Cutler, M. Comparative Assessment of Technologies for Extraction of Artemisinin. *J. Nat. Prod.* **2006**, *69*, 1653–1664. [CrossRef] [PubMed]
29. Cecchi, T. Biocascading: General Strategy for the Recovery of Valuable Substances from Food Waste. In *Biobased Products from Food Sector Waste: Bioplastics, Biocomposites, and Biocascading*; Cecchi, T., De Carolis, C., Eds.; Springer International Publishing: Cham, Switzerland, 2021; pp. 109–167. ISBN 978-3-030-63436-0.
30. Qi, X.; Tester, R.F. Lactose, Maltose, and Sucrose in Health and Disease. *Mol. Nutr. Food Res.* **2020**, *64*, 1901082. [CrossRef] [PubMed]
31. Zhang, M.; Ma, X.; Xiao, Z.; Sun, A.; Zhao, M.; Wang, Y.; Huang, D.; Sui, X.; Huo, J.; Zhang, Y. Polyphenols in Twenty Cultivars of Blue Honeysuckle (*Lonicera caerulea* L.): Profiling, Antioxidant Capacity, and α-Amylase Inhibitory Activity. *Food Chem.* **2023**, *421*, 136148. [CrossRef]
32. Khattab, R.; Brooks, M.S.-L.; Ghanem, A. Phenolic Analyses of Haskap Berries (*Lonicera caerulea* L.): Spectrophotometry Versus High Performance Liquid Chromatography. *Int. J. Food Prop.* **2016**, *19*, 1708–1725. [CrossRef]

33. Munteanu, I.G.; Apetrei, C. Analytical Methods Used in Determining Antioxidant Activity: A Review. *Int. J. Mol. Sci.* **2021**, *22*, 3380. [CrossRef] [PubMed]
34. Wang, C.; Gong, X.; Bo, A.; Zhang, L.; Zhang, M.; Zang, E.; Zhang, C.; Li, M. Iridoids: Research Advances in Their Phytochemistry, Biological Activities, and Pharmacokinetics. *Molecules* **2020**, *25*, 287. [CrossRef] [PubMed]
35. Kucharska, A.Z.; Sokół-Łętowska, A.; Oszmiański, J.; Piórecki, N.; Fecka, I. Iridoids, Phenolic Compounds and Antioxidant Activity of Edible Honeysuckle Berries (*Lonicera caerulea* Var. *Kamtschatica* Sevast.). *Molecules* **2017**, *22*, 405. [CrossRef] [PubMed]
36. Suomi, J.; Sirén, H.; Hartonen, K.; Riekkola, M.-L. Extraction of Iridoid Glycosides and Their Determination by Micellar Electrokinetic Capillary Chromatography. *J. Chromatogr. A* **2000**, *868*, 73–83. [CrossRef]
37. Turner, A.; Chen, S.-N.; Nikolic, D.; van Breemen, R.; Farnsworth, N.R.; Pauli, G.F. Coumaroyl Iridoids and a Depside from Cranberry (*Vaccinium macrocarpon*). *J. Nat. Prod.* **2007**, *70*, 253–258. [CrossRef] [PubMed]
38. Juadjur, A.; Winterhalter, P. Development of a Novel Adsorptive Membrane Chromatographic Method for the Fractionation of Polyphenols from Bilberry. *J. Agric. Food Chem.* **2012**, *60*, 2427–2433. [CrossRef] [PubMed]
39. Kucharska, A.Z.; Szumny, A.; Sokół-Łętowska, A.; Piórecki, N.; Klymenko, S.V. Iridoids and Anthocyanins in Cornelian Cherry (*Cornus mas* L.) Cultivars. *J. Food Compos. Anal.* **2015**, *40*, 95–102. [CrossRef]
40. Leisner, C.P.; Kamileen, M.O.; Conway, M.E.; O'Connor, S.E.; Buell, C.R. Differential Iridoid Production as Revealed by a Diversity Panel of 84 Cultivated and Wild Blueberry Species. *PLoS ONE* **2017**, *12*, e0179417. [CrossRef] [PubMed]
41. Heffels, P.; Müller, L.; Schieber, A.; Weber, F. Profiling of Iridoid Glycosides in *Vaccinium* Species by UHPLC-MS. *Food Res. Int.* **2017**, *100*, 462–468. [CrossRef] [PubMed]
42. Cacace, J.; Mazza, G. Optimization of Extraction of Anthocyanins from Black Currants with Aqueous Ethanol. *J. Food Sci.* **2003**, *68*, 240–248. [CrossRef]
43. Mohsen-Nia, M.; Amiri, H.; Jazi, B. Dielectric Constants of Water, Methanol, Ethanol, Butanol and Acetone: Measurement and Computational Study. *J. Solut. Chem.* **2010**, *39*, 701–708. [CrossRef]
44. Chen, X.-Q.; Li, Z.-H.; Wang, Z.-J.; Liu, L.-L.; Sun, T.-T.; Ma, J.-Z.; Zhang, Y. Ultrasound-Assisted Extraction of Total Anthocyanins from *Rubia sylvatica* Nakai Fruit and Radical Scavenging Activity of the Extract. *Ind. Crops Prod.* **2020**, *150*, 112420. [CrossRef]
45. Liao, J.; Xue, H.; Li, J. Extraction of Phenolics and Anthocyanins from Purple Eggplant Peels by Multi-Frequency Ultrasound: Effects of Different Extraction Factors and Optimization Using Uniform Design. *Ultrason. Sonochem.* **2022**, *90*, 106174. [CrossRef]
46. Lee, J.; Durst, R.W.; Wrolstad, R.E. Determination of Total Monomeric Anthocyanin Pigment Content of Fruit Juices, Beverages, Natural Colorants, and Wines by the Ph Differential Method: Collaborative Study. *J. AOAC Int.* **2005**, *88*, 1269–1278. [CrossRef] [PubMed]
47. Huber, G.M.; Rupasinghe, H.P.V. Phenolic Profiles and Antioxidant Properties of Apple Skin Extracts. *J. Food Sci.* **2009**, *74*, C693–C700. [CrossRef] [PubMed]
48. Benzie, I.F.; Strain, J.J. The Ferric Reducing Ability of Plasma (FRAP) as a Measure of "Antioxidant Power": The FRAP Assay. *Anal. Biochem.* **1996**, *239*, 70–76. [CrossRef]
49. Guo, L.; Qiao, J.; Mikhailovich, M.S.; Wang, L.; Chen, Y.; Ji, X.; She, H.; Zhang, L.; Zhang, Y.; Huo, J. Comprehensive Structural Analysis of Anthocyanins in Blue Honeysuckle (*Lonicera caerulea* L.), Bilberry (*Vaccinium uliginosum* L.), Cranberry (*Vaccinium macrocarpon* Ait.), and Antioxidant Capacity Comparison. *Food Chem. X* **2024**, *23*, 101734. [CrossRef]
50. Huang, D.; Ou, B.; Hampsch-Woodill, M.; Flanagan, J.A.; Prior, R.L. High-Throughput Assay of Oxygen Radical Absorbance Capacity (ORAC) Using a Multichannel Liquid Handling System Coupled with a Microplate Fluorescence Reader in 96-Well Format. *J. Agric. Food Chem.* **2002**, *50*, 4437–4444. [CrossRef] [PubMed]

molecules MDPI

Article

Characterization of Bioactive Phenolic Compounds Extracted from Hydro-Distillation By-Products of Spanish *Lamiaceae* Plants

Silvia Pérez-Magariño *, Marta Bueno-Herrera and M. Carmen Asensio-S.-Manzanera

Agrarian Technological Institute of Castilla and León, Consejería de Agricultura y Ganadería, Ctra Burgos Km 119, Finca Zamadueñas, 47071 Valladolid, Spain; bueherma@itacyl.es (M.B.-H.); asesanmr@itacyl.es (M.C.A.-S.-M.)
* Correspondence: permagsi@itacyl.es; Tel.: +34-983-41-52-45

Abstract: Plants of the *Lamiaceae* family are widely used for the extraction of essential oils, and this industry generates a large number of solid residues as by-products, which contain non-volatile valuable compounds. The aim of this work was to identify and quantify the phenolic compounds present in these solid residues from different important Spanish species of *Lamiaceae* to characterize and valorize them. Forty-seven phenolic compounds were identified by HPLC-DAD-MS and quantified by HPLC-DAD. Different concentrations and types of phenolic compounds were found between the solid residues. The *Rosmarinus officinalis* extracts showed the highest total phenolic content due to their high phenolic terpene concentrations. The *Thymus mastichina* extracts were characterized by kaempferol and flavanones, and some flavones were derived from luteolin and apigenin. Finally, the sample *Lavandula* and *Salvia lavandulifolia* extracts presented the lowest content of most phenolic compounds, with the exception of some phenolic acids, such as danshensu, salvianolic acid A, and glucosides of hydroxycinnamic acids. Therefore, this work provides information on the quantification of a large number of phenolic compounds using a simple, sensitive, reproducible, and accurate methodology. In addition, the results indicate that these solid residues still contain important amounts of different polyphenols, which are antioxidants and can be used in different industries.

Keywords: phenolic acids; flavonoids; terpenes; aromatic plants; solid residue

1. Introduction

Lamiaceae is a cosmopolitan family of great diversity, comprising a large number of species. Species belonging to this family are commonly used as medicinal and/or aromatic plants (MAPs) because of their ability to synthesize volatile compounds and also because they are an important source of polyphenols. Plants of the *Lamiaceae* family are mainly used as ornamentals for the extraction of essential oils (EOs), for the food industry as nutraceuticals, and for phytotherapy and have increasing value and potential worldwide [1–3]. The global market for EOs was valued at USD 7.6 billion in 2018 and is expected to reach USD 15.1 billion by the year 2026, as evidence of the growing use of these products emerges worldwide [4].

EOs are usually extracted from raw plant material (flowers, buds, seeds, leaves, twigs, bark, herbs, wood, fruits and roots) by hydro-distillation or by steam distillation.

The EO yield is 0.5–8 *w*/*w* % of the dry plant, so the industrial processing of MAPs generates a large number of solid residues as by-products. Between 2017 and 2021, this industry generated more than 250 thousand tons of distillation residues per year [5]. Their disposal can lead to serious environmental problems, especially if they are incinerated or landfilled. Nowadays, the main challenge for the MAP sector is to add value to the by-products of the distillation and drying process [2]. Some industries attempt to valorize such waste by using them for energy or composting [6]. These recycling systems have some disadvantages. The industry requires a huge investment to recycle the by-products into energy. On the other hand, recycling for composting is not always satisfactory due to the antigermination properties of some MAPs, which can also be transferred from the plant residues. There is an increasing interest in recovering valuable bioactive compounds from the residues of the EO industry for their valorization [7]. These solid distillation residues still contain non-volatile valuable compounds, such as different polyphenols, which can be extracted and used as antioxidants in the food, feed, or cosmetic industries [8,9].

It is known that the distillation process can significantly affect the presence and number of different phenols in the solid residues, but nevertheless, they are an important source of phenolic compounds [6,10–15]. These compounds can be extracted using several solvents and their mixtures, such as methanol, ethanol, acetone, ethyl acetate, or water [14]. The nature and content of phenolic compounds varies widely between aromatic plants and species. The most important compounds in the *Lamiaceae* family are phenolic acids, flavonoids, and phenolic terpenes [16–18]. Therefore, the use of organic solvents should be necessary to favor the extraction of more non-polar compounds, mainly flavonoids and phenolic terpenes. Liquid Chromatography with Diode Array Detection and/or Mass Spectrometry is the most commonly used technique to evaluate and characterize the phenolic compounds of different plant materials [17–19]. The identification and quantification of phenolic compounds in plant material is a difficult task due to the complexity of polyphenolic structures, the large number of compounds that can be found, and the limited number of commercially available standards.

Therefore, most of the studies focus on the tentative identification of a large number of compounds using complex techniques with MS/MS or Q-TOF-MS detectors, but only some of them are focused on the quantification of these compounds. Moreover, these techniques require a high level of investment, infrastructure, and highly skilled personnel, which are beyond the reach of most industries and laboratories. The aim of this work was to identify and quantify as many as possible of the phenolic compounds present in the solid residue obtained after the EO extraction from different important Spanish species of *Lamiaceae* in order to characterize and add value to these residues. Six different aromatic plants were studied: *Thymus mastichina* L. (TM), *Salvia lavandulifolia* Vahl (SL), *Lavandula angustifolia* Mill. (LA), *Lavandula latifolia* Medik (LL), three types of *Lavandula x intermedia* Emeric ex Loisel. (LI_ Abrial, LI_Grosso and LI_Super), and *Rosmarinus officinalis* L. (RO). In addition, a simple analytical method using High-Performance Liquid Chromatography with Diode Array Detection (HPLC-DAD) and Mass Spectrometry (MS) was developed and validated. This work provides a simple methodology that allows, for the first time, the complete characterization and simultaneous quantification of 47 phenolic compounds in the by-products obtained from the distillation process of different Spanish MAPs. It has also highlighted the important amounts of bioactive phenolic compounds that can be extracted from these MAPs for use in different industries.

2. Results and Discussion

2.1. Identification of Phenolic Compounds in the Solid Residue from Hydro-Distillation of the Aromatic Plants

The extraction technique and the chromatographic method used allowed the characterization of forty-seven phenolic compounds, grouped into phenolic acids, flavonoids (flavones, flavanones, and flavonols), and phenolic terpenes. Table 1 summarizes the spectroscopic and the spectrometric data of all the identified phenolic compounds, numbered according to their elution order. Figure 1 shows the chromatograms of the different phenolic profiles of the *Lamiaceae* plants studied at two different wavelengths.

Table 1. Peak number, spectroscopic and spectrometric data, and quantification wavelength (QW) of the phenolic compounds identified in the solid residue of the EO by-product of *Lamiaceae* aromatic plants.

Peak	Compound	Phenolic Class	λ Max (nm)	[M-H]⁻	Main Fragments	QW (nm)	IdC [b]	TM [c]	SL [c]	L [c]	RO [c]
1	Danshensu (3,4-dihydroxyphenyl lactic acid)	Phenolic acid	282	197	179, 135	280	S		X [e]	X	X
2	Chlorogenic acid	Phenolic acid	300sh [a], 326	353	191, 179	320	S	X			X
3	Cryptochlorogenic acid	Phenolic acid	300sh, 326	353	191, 179	320	T				X
4	Coumaric acid-O-glucoside	Phenolic acid	264, 296sh	325	163, 119	254	T			X	
5	p-hydroxybenzoic acid	Phenolic acid	256	137		254	S	X			X
6	Ferulic acid-O-glucoside 1	Phenolic acid	282sh, 304	355	193, 149	320	T			X	
7	Caffeic acid	Phenolic acid	298sh, 320	179	135	320	S	X	X	X	X
8	Dihydro-p-coumaric acid glucoside	Phenolic acid	276, 314sh	327	165, 121	280	T			X	
9	6-hydroxyluteolin-7-O-glucoside	Flavone	282, 344	463	301	350	T	X			X
10	Ferulic acid-O-glucoside 2	Phenolic acid	294, 320	355	193, 149	320	T			X	
11	Rosmarinic acid-3-O-glucoside	Phenolic acid	288sh, 322	521	359	320	T				X
12	Luteolin-7-O-rutinoside	Flavone	254, 266, 350	593	285	350	S		X	X	X
13	Luteolin-7-O-glucuronide	Flavone	258, 266, 350	461	285	350	S				X
14	Luteolin-7-O-glucoside	Flavone	258, 268, 350	447	285	350	S	X	X	X	X
15	Naringenin-7-O-glucoside	Flavanone	284, 332sh	433	271	280	S	X			
16	Hesperetin-7-rutinoside (hesperidin)	Flavanone	284, 334sh	609	301	280	S		X		X
17	Nepetin-7-O-glucoside (nepitrin)	Flavone	258, 272, 348	477	315	350	S				X
18	Rosmarinic acid	Phenolic acid	290sh, 330	359	197, 179, 161	320	S	X	X	X	X
19	Apigenin-7-O-glucoside	Flavone	268, 336	431	269	320	S	X	X	X	
20	6-hydroxyluteolin	Flavone	252–268, 350	301		350	T	X			
21	Salvianolic acid A	Phenolic acid	288, 310sh, 336sh	493	295	280	S		X	X	
22	Hispidulin-7-O-glucoside	Flavone	276, 334	461	299	350	T				X
23	Eriodictyol	Flavanone	288, 334sh	287	151, 135	280	S	X			
24	Scutellarein-7-O-glucuronide	Flavone	270, 338	461	285	350	T				X
25	Scutellarein	Flavone	286, 338	285		350	S	X			
26	Luteolin-3′-acetyl-O-glucuronide	Flavone	270, 338	503	399, 285	350	T				X
27	Naringenin	Flavanone	290, 332sh	271		280	S	X			
28	Luteolin	Flavone	254, 266, 350	285		350	S	X	X	X	X
29	Kaempferol	Flavonol	266, 366	285		350	S	X			
30	Cirsiliol	Flavone	256–274, 346	329		350	S		X		X
31	Rosmanol	Phenolic terpene	290	345	283	280	S				X
32	Apigenin	Flavone	268, 340	269		350	S	X	X	X	X
33	Eupatorin	Flavone	276, 344	345 [d]		350	S		X		
34	Cirsimaritin	Flavone	274, 336	313	298, 283	350	S		X		X

Table 1. *Cont.*

Peak	Compound	Phenolic Class	λ Max (nm)	$[M-H]^-$	Main Fragments	QW (nm)	IdC [b]	TM [c]	SL [c]	L [c]	RO [c]
35	Ladanein	Flavone	284, 334	315 [d]		350	T		X	X	X
36	Sakuranetin	Flavanone	288, 330sh	285		280	S	X			
37	Acacetin	Flavone	268, 334	283		350	S				X
38	Genkwanin	Flavone	268, 338	283		350	S				X
39	Salvigenin	Flavone	276, 332	329 [d]	287 [d]	320	S		X		X
40	(Epi)rosmanol methyl ether	Phenolic terpene	290	359	283	280	T				X
41	Carnosol	Phenolic terpene	284	329	285	280	S		X		X
42	Rosmadial	Phenolic terpene	284	343	299, 285	280	T				X
43	4′-methoxytectochrysin	Flavone	268, 332	299 [d]		350	T				X
44	Carnosol isomer	Phenolic terpene	284	329	285	280	T				X
45	Carnosic acid derivative	Phenolic terpene	280	301		280	T				X
46	Carnosic acid	Phenolic terpene	286	331	287	280	S		X		X
47	12-methylcarnosic acid	Phenolic terpene	284	345	301, 286	280	T				X

[a] sh: shoulder. [b] IdC: identification compound. S: standard compound; T: tentatively identified compound with data reported in the literature. [c] TM: *Thymus mastichina;* SL: *Salvia lavandulifolia;* L: *Lavandula* spp.; RO: *Rosmarinus officinalis.* [d] $[M-H]^+$. [e] compounds that appear in the different MAPs.

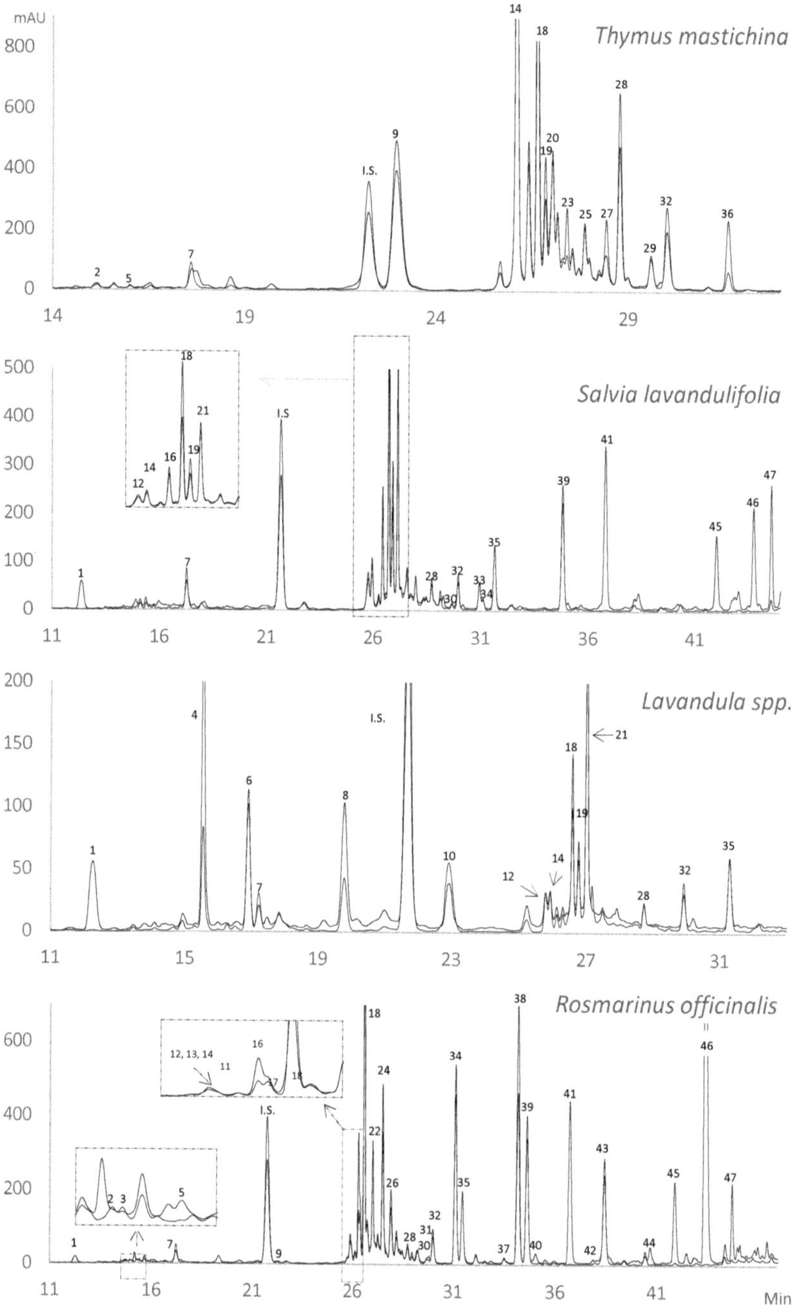

Figure 1. Chromatograms of the phenolic compounds of *Thymus mastichina*, *Salvia lavandulifolia*, *Lavandula* spp., and *Rosmarinus officinalis* monitored at 280 (dotted line) and 320 nm (continuous line). Peak numbers are the compounds shown in Table 1. IS: internal standard.

Twenty-nine compounds were identified by the standard compounds, and eighteen were tentatively identified from their UV absorption spectra and mass spectral data reported in the bibliography, as discussed below.

2.1.1. Phenolic Acids

Twelve phenolic acids were identified, six by standard compounds (danshensu (3,4-dihydroxyphenyl lactic acid), chlorogenic acid, p-hydroxybenzoic acid, caffeic acid, rosmarinic acid, and salvianolic acid A) and another six were tentatively identified. Peak 3 had the same UV absorption spectrum as chlorogenic acid (UVmax 326 nm), with the same *m*/*z* 353 and fragments. It was identified as crytochlorogenic acid, which was also supported by Irakli et al. [18], who identified this compound by its standard and it was detected in RO.

Peak 4 presented the molecular ion *m*/*z* 325 and a fragment ion at 163 *m*/*z* corresponding to the coumaric acid and a loss of 162 amu that reveals the presence of glucose or galactose. However, no galactoside compounds of hydroxycinnamic acids have been found in the literature, so this glycoside is most probably glucoside [20], and therefore, this compound was tentatively identified as coumaric acid-O-glucoside [8,17]. Similarly, peaks 6 and 10 were identified as glucosides of ferulic acid since they presented the molecular ion at *m*/*z* 355 and fragment ions at *m*/*z* 193, corresponding to the ferulic acid with a loss of glucose. These two compounds were also identified by Torras-Claveria et al. [8] and Contreras et al. [17] in the genus *Lavandula* and other *Lamiaceae* species, respectively. Other glucosides of hydroxycinnamic acids identified were dihydro-p-coumaric acid glucoside or hydroxyhydrocinnamic acid glucoside (peak 8) found in the genus *Lavandula* and rosmarinic acid-3-O-glucoside (peak 11) in RO. The mass fragmentation data of peak 8 showed the molecular ion at *m*/*z* 327 and a fragment ion at *m*/*z* 165, indicating the loss of glucose.

These data agree with those found by Contreras et al. [17], who identified this compound in *Helleborus niger*. Rosmarinic acid-3-O-glucoside presented the molecular ion *m*/*z* 521, and it was also identified by Borras-Linares et al. [21] and Sharma et al. [22] in RO.

2.1.2. Flavones and Flavonols

Twenty-one flavones were identified, of which fourteen were standard compounds (luteolin-7-O-rutinoside, luteolin-7-O-glucuronide, luteolin-7-O-glucoside, nepetin-7-O-glucoside (nepitrin), apigenin-7-O-glucoside, scutellarein, luteolin, cirsiliol, apigenin, eupatorine, cirsimaritin, acacetin, genkwanin and salvigenin) and seven were tentatively identified. All of the flavones were derivatives of apigenin and luteolin. Peaks 9, 20, and 26 were luteolin derivatives. Peak 9 was identified as 6-hydroxyluteolin-7-O-glucoside and presented molecular ions at *m*/*z* 463 and a fragment ion at *m*/*z* 301 due to the loss of a glucose moiety. This compound was found in TM and RO and was also identified by Borras-Linares et al. [21] and Sharma et al. [22] in RO. Peak 20 presented the ion *m*/*z* 301 and the same UV absorption spectrum as luteolin, and it was tentatively identified as 6-hydroxyluteolin, being identified only in TM. Peak 26 had a molecular ion at *m*/*z* 503 and fragment ions at *m*/*z* 399 and 285. This fragmentation pattern agreed with one of the isomers of luteolin-3′-O-(O-acetyl)-β-D-glucuronide identified in RO by Borras-Linares et al. [21], Mena et al. [16], Achour et al. [23], and Sharma et al. [22] and it could be luteolin-3′-O-(2″-O-acetyl)-β-D-glucuronide [21,23].

Peaks 22, 24, 35, and 43 were apigenin derivatives, all of which were identified in RO, and peak 35 was also identified in SL and the genus *Lavandula*. Peaks 22 and 24 presented the molecular ions at *m*/*z* 461 and different fragment ions at *m*/*z* 299 and 285, respectively.

The first peak (peak 22) was identified as hispidulin-7-glucoside, with the fragment ion *m*/*z* 299 corresponding to hispidulin with a loss of glucose (162 amu), also called homoplantaginin according to the data found in the bibliography [18,22,24].

Peak 24 was identified as scutellarin or scutellarein-7-glucuronide, with the fragment ion *m*/*z* 285 corresponding to scutellarein with a loss of glucuronide acid (176 amu), as was previously reported by Sharma et al. [22].

The molecular ion of peak 35 was only found in the positive mode with m/z 315, which may correspond to ladanein, a dimethyl scutellarein derivative, with methylations at 4′ and 7 hydroxyls [25,26]. According to Psarrou et al. [26], this compound showed two maximum absorbances at 286 and 334 nm, which agrees with the data in our study (Table 1). This pattern of the protonated molecular ion [M-H] + was also found in two other flavones, eupatorin and salvigenin as compounds for which a standard was available. Peak 43 also presented an apigenin-type UV spectrum with maximum absorbances at 268 and 332 nm with a molecular ion at m/z 299 in the positive mode. This compound was identified as 4′-methoxytectochrysin or apigenin-4′,7-dimethylether, taking into account the data found in the bibliography and the fact that it was the only flavonoid that elutes between carnosol and carnosic acid [19,25–27].

Five flavanones were identified by their respective standards, naringenin, naringenin-7-O-glucoside, eriodictyol, sakuranetin, and hesperidin. The first four compounds were only detected in the TM extracts, and hesperidin was only detected in the SL and RO extracts.

Kaempferol was identified by its standard, and it was the only flavonol detected in the TM extracts.

2.1.3. Phenolic Terpenes

Finally, the phenolic terpenes were detected in the solid residue of RO and SL. Carnosic acid is known to be unstable, and its degradation is favored by temperature, exposure to light, and the presence of oxygen. Under these conditions, carnosic acid can be oxidized to carnosol, rosmanol, epirosmanol, epiisorosmanol, methylepirosmanol, rosmadial, and methylcarnosate, among others [28–30]. Bicchi et al. [31] and Troncoso et al. [32] reported that carnosol is the main compound obtained after the degradation of carnosic acid, which can be transformed into other phenolic diterpenes such as rosmanol and its isomers, epirosmanol, and epiisorosmanol. In addition, the use of methanol in the extraction and under acid conditions could break the lactone bond of carnosol, giving rise to other compounds derived from carnosic acid, and carnosol [29].

In this study, eight phenolic terpenes were identified: three by standards (rosmanol, carnosol and carnosic acid), and the other five were tentatively identified. All these compounds showed similar UV spectra to carnosic acid and carnosol with maximum absorbances at 284 nm, so they are derived from carnosic acid.

Peak 40 presented the molecular ion at m/z 359 and a fragment ion at m/z 283, corresponding to the loss of a methoxy (-O-CH_3) and an acid (-COOH) group. This compound was identified as rosmanol methyl ether or epirosmanol methyl ether according to Borras-Linares et al. [21], Mena et al. [16], and Sharma et al. [22]. This compound can be formed from carnosol due to the use of methanol for extraction, which agrees with Doolaege et al. [29]. Zhang et al. [30] obtained the corresponding ethyl ether of epirosmanol as a degradation product of carnosol when carnosol was dissolved in ethanol. These degradation products were not formed when an aprotic solvent such as acetonitrile was used.

Peak 42 had a molecular ion at m/z 343 and a fragment ion at m/z 299 due to the loss of an acid group (-COOH), and a maximum absorbance at 288 nm. These data agreed with those found in the bibliography [21,22,30] and it was identified as rosmadial.

Peaks 44 and 47 also presented UV absorption spectra similar to carnosol and carnosic acid and molecular ions at m/z 329 and 345, respectively, which were identified as carnosol isomer and 12-methylcarnosic acid, according to other studies [21,22]. The carnosol isomer showed the same fragmentation pattern as carnosol (with molecular ions at m/z 329 and a fragment ion at m/z 285). On the other hand, 12-methylcarnosic acid had two fragment ions (m/z 301 and 286) corresponding to the loss of the acid group (44 amu) and the subsequent loss of the methyl group (15 amu).

Finally, peak 45 presented a molecular ion at m/z 301 and showed a UV absorption maximum at 280 nm. Different authors have reported a compound with these mass and UV spectral characteristics, but they have assigned different structures. Borras-Linares et al. [33]

identified it as salviol, which is an abietatriene compound with a single phenol group isolated from some Salvia species. Doolaege et al. [29] identified and characterized one compound with a molecular weight of 302, which is a fragment ion of m/z 301 and has a UV spectrum similar to carnosic acid. They hypothesized that this compound results from the breakdown of the intramolecular bond of rosmanol and its isomers under acid conditions, followed by the formation of an unsaturated double bound and the loss of a CO-group. The UV spectrum of this compound had an absorbance pattern similar to that of our peak 45. Finally, Zhang et al. [30] proposed a structure that was named 5,6,7,10-tetrahydro-7-hydroxy-rosmariquinone, which was a major degradation product derived from carnosic acid. In all these studies, as in our work, this compound eluted before carnosic acid. With the data available in this study, we could not confirm the structure of this compound, but it was clear that it was derived from carnosic acid. Therefore, this compound has been named a carnosic acid derivative.

Finally, two rosmanol isomers, epiisorosmanol, and epirosmanol, were identified by their MS fragmentation patterns. They presented the same molecular ion at m/z 345 as rosmanol, but they had some differences in the fragmentation pattern and eluted at different retention times. According to Kontogianni et al. [24], Achour et al. [23], and Sharma et al. [22], the first isomer, epiisorosmanol, eluted before rosmanol and showed the major fragment ions at m/z 301 and 283, corresponding to the loss of 44 amu (acid group) and later of 18 amu (water molecule). The second, epirosmanol, eluted after rosmanol and showed only the major fragment at m/z 283. These isomers were clearly identified by MS. However, they were not quantified due to the low signal found in the DAD.

2.2. Results of the Method Validation

Table 2 shows the range, calibration curve, correlation coefficients, LOD, and LOQ for each standard phenolic compound. The linearity in the range indicated that all compounds were satisfactory, with correlation coefficients between 0.9952 and 0.9999. The LOD and LOQ values indicated that the developed method had sufficient sensitivity for the analysis of phenolic compounds in these aromatic plant residues.

The repeatability and reproducibility values were satisfactory, ranging from 0.8 to 7.3% for repeatability and from 2.3 to 9.4% for reproducibility, being less than 10% for all the phenolic compounds (Table 3).

In general, good recoveries were obtained, ranging from 80 to 120% for most of the compounds. Only kaempferol, carnosol, and carnosic acid showed recoveries outside this range. Kaempferol showed recoveries of around 65%, which may be due to the fact that this compound is very reactive and may react with other phenolic compounds [34]. Similarly, carnosol and carnosic acid showed recoveries of around 135% and 67%, respectively. This may be due to the fact that carnosic acid is unstable and can be oxidized to carnosol and other related compounds [28–30], as previously commented. Furthermore, carnosol is the main compound obtained after carnosic acid degradation [31,32], which is consistent with the loss of about 33% of carnosic acid and the increase to 33% of carnosol.

Therefore, all these results indicate that this method can be considered precise, reproducible, and accurate.

2.3. Quantification of Phenolic Compounds in the Solid Residue from Hydro-Distillation of the Aromatic Plants

Table 4 shows the concentration of the phenolic compounds quantified in the solid residue of the EO by-product of the different *Lamiaceae* aromatic plants studied. The methodology described above allowed the identification and quantification of forty-seven phenolic compounds. However, not all the compounds were found in the solid residue of the different plants. Only five compounds, caffeic acid, rosmarinic acid, luteolin-7-O-glucoside, luteolin, and apigenin, were found in all the studied by-products.

Table 2. Range, linearity, correlation coefficient (R^2), sensibility data, and accuracy of the standard phenolic compounds.

Compound	Range (μg/mL)	Calibration [a]	R^2	LOD [b] (μg/mL)	LOQ [b] (μg/mL)	Supplier [c]	Amounts Added (μg/mL)	Recovery (% RSD, *n* = 6)	
								TM	SL
Danshensu	2.0–120	y = 0.0044x + 0.0012	0.9984	1.1	4.3	1	24/48	-	97/90
Chlorogenic acid	1.0–10	y = 0.0130x − 0.0006	0.9973	0.37	1.1	1	1/2	102/100	
p-hydroxybenzoic acid	0.94–28	y = 0.0427x + 0.0129	0.9996	0.07	0.92	2	1/2	94/94	
Caffeic acid	1.0–31	y = 0.0223x − 0.0010	0.9987	0.70	2.2	1	4/8	105/106	120/112
Luteolin-7-O-rutinoside	1.3–60	y = 0.0093x − 0.0009	0.9997	0.78	2.3	1	4/8		102/95
Luteolin-7-O-glucuronide	2.0–160	y = 0.0122x + 0.0224	0.9991	0.33	3.4	3			
Naringenin-7-O-glucoside	2.0–40	y = 0.0155x + 0.0063	0.9998	0.11	1.3	4			
Rosmarinic acid	11–123	y = 0.0104x − 0.0268	0.9998	3.7	6.3	1	30/60	80/81	116/98
Apigenin-7-O-glucoside	2.1–83	y = 0.0120x + 0.013	0.9993	0.57	0.95	4			
Salvianolic acid A	2.0–80	y = 0.0160x + 0.0136	0.9977	0.96	5.2	1	6/12		118/102
Eriodictyol	2.1–52	y = 0.0212x + 0.0088	0.9996	0.39	2.3	4			
Scutellarein	0.98–39	y = 0.0212x + 0.0009	0.9986	0.66	2.3	4			
Naringenin	2.0–40	y = 0.0230x + 0.0102	0.9997	0.32	1.8	4			
Luteolin	3.1–78	y = 0.0200x + 0.0064	0.9997	0.38	2.0	1			
Cirsiliol	0.42–2.8	y = 0.0285x − 0.0020	0.9974	0.14	0.30	1			
Kaempferol	1.0–40	y = 0.0130x − 0.0010	0.9999	0.26	0.68	2	7/14	60/65	
Rosmanol	2.2–174	y = 0.0018x − 0.0009	0.9994	0.89	4.1	3			
Apigenin	1.1–46	y = 0.0175x − 0.0060	0.9997	0.35	0.82	1	3/6	94/96	108/103
Cirsimaritin	1.3–60	y = 0.0072x + 0.0004	0.9998	0.45	1.6	1			
Eupatorin	0.96–19	y = 0.0226x − 0.0011	0.9996	0.23	0.65	4	4/8		110/116
Sakuranetin	2–40	y = 0.0201x − 0.0067	0.9999	0.18	1.4	4			
Salvigenin	1.0–60	y = 0.0147x − 0.0247	0.9952	3.7	7.3	5			
Carnosol	2.0–200	y = 0.0023x + 0.0038	0.9985	0.69	3.5	4	30/60		140/135
Carnosic acid	10–200	y = 0.0010x − 0.0062	0.9974	10	19	4	80/160		64/67
Ferulic acid	3.9–90	y = 0.0254x − 0.0046	0.9998	0.54	2.2	2			
trans-cinnamic acid	5.9–59	y = 0.0564x − 0.0094	0.9997	0.10	0.71	2			

[a] y: relative area; x: concentration (μg/mL). [b] LOD: limit of detection; LOQ: limit of quantification. [c] 1: Sigma–Aldrich (Steinheim, Germany); 2: Fluka (Buchs, Switzerland); 3: Phytolab (Vestenbergsgreuth, Germany); 4: Extrasynthèse (Lyon, France); 5: Targetmol (Wellesley Hills, MA, USA).

Table 3. Precision for the phenolic compounds quantified in the solid residue of the EO by-product of *Lamiaceae* aromatic plants studied.

	Repetibility (%RSD, n = 3)				Reproducibility (%RSD, *n* = 9)			
Phenolic Compounds	**TM**	**SL**	**LL**	**RO**	**TM**	**SL**	**LL**	**RO**
Phenolic acids								
Danshensu	-	5.0	2.9	5.3	-	6.4	7.4	5.4
Chlorogenic acid	2.2	-	-	2.9	6.5	-	-	3.5
Cryptochlorogenic acid	-	-	-	3.4	-	-	-	6.5
Coumaric acid-O-glucoside	-	-	2.6	-	-	-	8.1	-
p-hydroxybenzoic acid	1.2	-	-	6.1	5.6	-	-	7.6
Ferulic acid-O-glucoside 1	-	-	2.6	-	-	-	7.6	-
Caffeic acid	1.6	2.6	1.0	1.6	3.7	4.7	6.4	3.0
Dihydro-p-coumaric acid glucoside	-	-	6.4	-	-	-	8.1	-
Ferulic acid-O-glucoside 2	-	-	2.2	-	-	-	7.2	-
Rosmarinic acid-3-O-glucoside	-	-	-	4.6	-	-	-	5.4
Rosmarinic acid	2.4	2.1	3.5	2.1	6.4	5.1	7.4	2.3
Salvianolic acid A	-	2.7	2.0	-	-	6.7	6.6	-
Flavones								
6-hydroxyluteolin-7-O-glucoside	2.2	-	-	4.0	5.8	-	-	6.2
Luteolin-7-O-rutinoside	-	2.9	3.8	5.5	-	5.5	5.8	6.2
Luteolin-7-O-glucuronide	-	-	-	4.1	-	-	-	7.6
Luteolin-7-O-glucoside	1.7	3.5	3.3	1.8	6.2	7.5	7.3	4.8
Nepetin 7-O-glucoside	-	-	-	2.0	-	-	-	5.0
Apigenin-7-O-glucoside	1.6	3.8	3.5	-	6.8	6.7	9.3	-
6-hydroxyluteolin	1.1	-	-	-	6.8	-	-	-
Hispidulin-7-O-glucoside	-	-	-	5.3	-	-	-	6.9
6-hydroxyapigenin-7-O-β-glucoside	1.4	-	-	-	7.0	-	-	-
Scutellarein-7-O-glucuronide	-	-	-	2.9	-	-	-	5.2
Scutellarein	3.6	-	-	-	5.9	-	-	-
Luteolin-3′-acetyl-O-glucuronide	-	-	-	2.2	-	-	-	5.2
Luteolin	0.8	3.8	3.5	2.6	4.6	5.1	5.6	3.6
Cirsiliol	-	5.9	-	1.9	-	6.1	-	3.4
Apigenin	0.9	5.4	4.3	1.5	3.7	7.3	8.7	3.1
Eupatorin	-	2.4	-	-	-	4.3	-	-
Cirsimaritin	-	2.1	-	1.5	-	3.9	-	2.6
Ladanein	-	5.1	5.8	1.7	-	5.2	8.7	3.1
Acacetin	-	-	-	2.6	-	-	-	5.5
Genkwanin	-	-	-	3.0	-	-	-	3.2
Salvigenin	-	3.5	-	2.7	-	4.7	-	3.2
4′-methoxytectochrysin	-	-	-	4.5	-	-	-	4.6
Flavanones								
Naringenin-7-O-glucoside	<LQ	-	-	-	<LQ	-	-	-
Hesperidin	-	6.2	-	6.0	-	9.3	-	8.2
Eriodictyol	3.2	-	-	-	5.8	-	-	-
Naringenin	2.6	-	-	-	3.8	-	-	-
Sakuranetin	4.5	-	-	-	5.6	-	-	-
Flavonols								
Kaempferol	1.3	-	-	-	4.0	-	-	-
Phenolic terpene								
Rosmanol	-	-	-	7.3	-	-	-	7.6
(Epi)rosmanol methyl ether	-	-	-	5.5	-	-	-	9.4
Rosmadial	-	-	-	4.7	-	-	-	7.1
Carnosol	-	3.1	-	2.7	-	4.1	-	2.8
Carnosol isomer	-	-	-	3.4	-	-	-	4.0
Carnosic acid derivative	-	-	-	5.3	-	-	-	6.7
Carnosic acid	-	<LQ	-	3.3	-	<LQ	-	5.3
12-methylcarnosic acid	-	-	-	4.0	-	-	-	5.1

LQ: limit of quantification.

Table 4. Phenolic compounds quantified in the solid residue of the EO by-product of *Lamiaceae* aromatic plants studied (μg/g of dry weight) [a].

Phenolic Compounds	RO	TM	SL	LL	LA	LI_Abrial	LI_Grosso	LI_Super
Phenolic acids								
Danshensu	307 ± 55	-	679 ± 285	1051 ± 451	593 ± 123	879 ± 311	457 ± 146	678 ± 254
Chlorogenic acid	44 ± 34	37 ± 17	-	-	-	-	-	-
Cryptochlorogenic acid	19 ± 7	-	-	-	-	-	-	-
Coumaric acid-O-glucoside	-	-	-	412 ± 163	100 ± 32	161 ± 50	156 ± 33	195 ± 28
p-hydroxybenzoic acid	16 ± 10	35 ± 18	-	-	-	-	-	-
Ferulic acid-O-glucoside 1	-	-	-	66 ± 30	131 ± 44	167 ± 58	132 ± 29	147 ± 28
Caffeic acid	88 ± 14	108 ± 32	123 ± 31	54 ± 19	79 ± 14	63 ± 18	114 ± 46	93 ± 54
Dihydro-p-coumaric acid glucoside	-	-	-	244 ± 97	69 ± 12	116 ± 37	107 ± 20	104 ± 22
Ferulic acid-O-glucoside 2	-	-	-	57 ± 29	158 ± 54	193 ± 75	164 ± 32	154 ± 37
Rosmarinic acid-3-O-glucoside	421 ± 284	-	-	-	-	-	-	-
Rosmarinic acid	2283 ± 720	1997 ± 1158	1486 ± 670	599 ± 272	579 ± 173	343 ± 109	235 ± 70	290 ± 67
Salvianolic acid A	-	-	457 ± 273	596 ± 304	252 ± 98	285 ± 135	282 ± 98	385 ± 113
Flavones								
6-hydroxyluteolin-7-O-glucoside	40 ± 38	1600 ± 751	-	-	-	-	-	-
Luteolin-7-O-rutinoside	61 ± 63	-	298 ± 161	158 ± 75	91 ± 20	150 ± 61	122 ± 23	146 ± 32
Luteolin-7-O-glucuronide	99 ± 91	-	-	-	-	-	-	-
Luteolin-7-O-glucoside	79 ± 67	2235 ± 903	195 ± 84	385 ± 118	14 ± 13	52 ± 19	45 ± 13	62 ± 20
Nepetin 7-O-glucoside	352 ± 190	-	-	-	-	-	-	-
Apigenin-7-O-glucoside	-	793 ± 285	426 ± 179	273 ± 70	153 ± 21	255 ± 98	162 ± 32	133 ± 72
6-hydroxyluteolin	-	440 ± 242	-	-	-	-	-	-
Hispidulin-7-O-glucoside	302 ± 62	-	-	-	-	-	-	-
6-hydroxyapigenin-7-O-β-glucoside	-	215 ± 75	-	-	-	-	-	-
Scutellarein-7-O-glucuronide	584 ± 151	-	-	-	-	-	-	-
Scutellarein	-	201 ± 155	-	-	-	-	-	-
Luteolin-3′-acetyl-O-glucuronide	364 ± 122	-	-	-	-	-	-	-
Luteolin	131 ± 43	636 ± 323	157 ± 52	42 ± 24	30 ± 10	49 ± 22	64 ± 17	60 ± 22
Cirsiliol	56 ± 37	-	20 ± 14	-	-	-	-	-
Apigenin	195 ± 82	665 ± 233	145 ± 61	29 ± 10	95 ± 35	100 ± 44	81 ± 21	64 ± 18
Eupatorin	-	-	95 ± 48	-	-	-	-	-
Cirsimaritin	1032 ± 509	-	382 ± 328	-	-	-	-	-
Ladanein	234 ± 141	-	74 ± 71	46 ± 14	93 ± 33	72 ± 20	93 ± 25	70 ± 18
Acacetin	127 ± 100	-	-	-	-	-	-	-
Genkwanin	606 ± 339	-	-	-	-	-	-	-
Salvigenin	398 ± 239	-	531 ± 343	-	-	-	-	-
4′-methoxytectochrysin	692 ± 308	-	-	-	-	-	-	-

Table 4. *Cont.*

Phenolic Compounds	RO	TM	SL	LL	LA	LI_Abrial	LI_Grosso	LI_Super
Flavanones								
Naringenin-7-O-glucoside	-	87 ± 85	-	-	-	-	-	-
Hesperidin	380 ± 148	-	41 ± 86	-	-	-	-	-
Eriodictyol	-	311 ± 132	-	-	-	-	-	-
Naringenin	-	310 ± 94	-	-	-	-	-	-
Sakuranetin	-	279 ± 103	-	-	-	-	-	-
Flavonols								
Kaempferol	-	268 ± 147	-	-	-	-	-	-
Phenolic terpene								
Rosmanol	218 ± 190	-	-	-	-	-	-	-
(Epi)rosmanol methyl ether	657 ± 434	-	-	-	-	-	-	-
Rosmadial	189 ± 205	-	-	-	-	-	-	-
Carnosol	6862 ± 1511	-	1055 ± 2203	-	-	-	-	-
Carnosol isomer	1809 ± 1605	-	-	-	-	-	-	-
Carnosic acid derivative	6937 ± 2416	-	559 ± 1703	-	-	-	-	-
Carnosic acid	26,703 ± 9259	-	867 ± 2687	-	-	-	-	-
12-Methylcarnosic acid	4174 ± 1789	-	1769 ± 4541	-	-	-	-	-
Total phenolic compounds	56,459 ± 1375	10,002 ± 3193	9301 ± 11,089	4015 ± 1343	2436 ± 463	2884 ± 859	2213 ± 341	2582 ± 431

[a] means ± standard deviation.

High differences were found between the solid residues from the *Lamiaceae* aromatic plants evaluated. The RO extracts showed the highest total phenolic content (56,459 ± 1375 μg/g) due to the high phenolic terpene concentrations. The extracts from TM and SL showed a medium and similar concentration of phenolics (10,002 ± 3193 μg/g and 9301 ± 11,089 μg/g, respectively), although the type of phenolics present in each plant was different. The TM extracts did not contain any phenolic terpenes and were characterized by a higher content of flavones and flavanones in comparison to the SL extracts. The SL extracts were mainly characterized by the presence of phenolic terpenes (46%), followed by phenolic acids (30%) and flavones (24%). Finally, the extracts from *Lavandula* spp. had the lowest total phenolic content of all the extracts studied (average of 2825 μg/g) and contained only phenolic acids and flavones.

It is noteworthy that differences in the content of phenolic compounds were found between the samples (Figure 2). Although all samples were collected in the same locality, they belonged to several seasons, and different weather conditions can affect the chemical composition. Furthermore, the variability was higher in wild species (TM, SL, LL, and RO) than in those that come from selected clones (*Lavandula* spp.) (Table 4, Figure 2). Variations in LA and lavandin were exclusively due to the weather conditions between years, while different populations from different locations were used to set the cultivation trials in wild species.

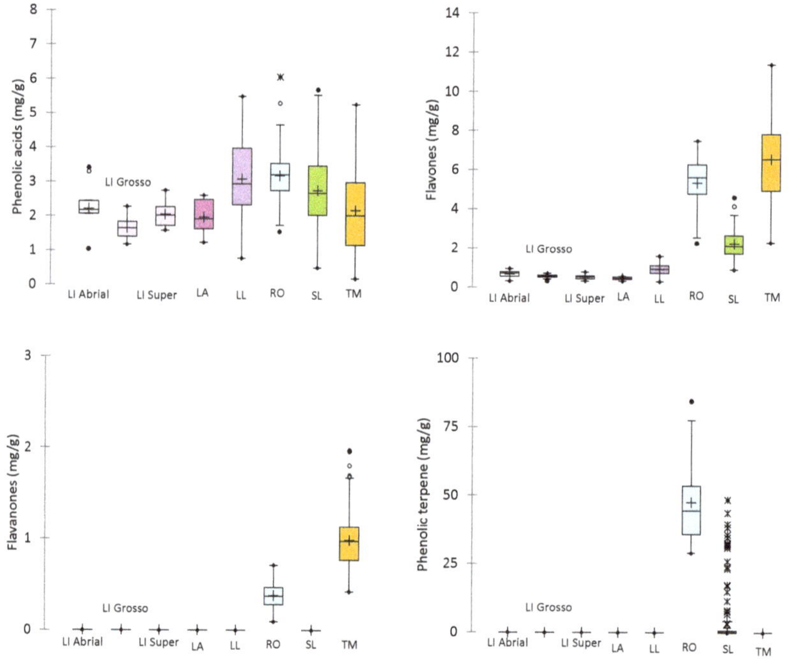

Figure 2. Box and Whisker plot of the different phenolic families of the solid residues from different Spanish species of *Lamiaceae*. +: mean values; •, ○ and ✳: outliers.

Rosmarinic acid was the major phenolic acid present in the aromatic plants studied as an ester obtained from the esterification product of 3,4-dihydroxyphenyl lactic acid (danshensu) and caffeic acid [35]. The extracts from RO and TM were the richest in rosmarinic acid, followed by the SL extracts. On the other hand, the solid residue of *Lavandula* spp presented the lowest concentration of rosmarinic acid (between 3 and 10 times lower than the content of the other plant extracts studied) but the highest concentration of danshensu, which is the precursor of rosmarinic acid. Salvianolic acid A was found in SL and *Lavandula* spp. extracts, which are formed by one molecule of danshensu and two molecules of caffeic acid [36]. Different salvianolic acids were found in high concentrations in *Salvia miltiorrhiza* (Danshen), which is a plant extensively used in traditional Chinese medicine for the treatment of cardiovascular diseases due to its antioxidant and free radical scavenging activities [37,38]. The rosmarinic acid glucoside was only found in the RO extract and is characteristic of this plant [21]. The hydroxycinnamic acid derivatives of coumaric and ferulic acids were only detected in *Lavandula* spp. extracts.

The flavonoids were the predominant compounds quantified in TM extracts at 78%. Among the flavonoids, the flavones were the most abundant phenolic group (66%), mainly formed by the aglicones luteolin and apigenin, 6-hydroxyluteolin, 6-hydroxyapigenin, and their O-glycoside derivatives. On the other hand, the RO extracts were characterized by the methyl-ether, methoxy, and glucuronide derivatives of luteolin and apigenin. Among the flavones, the SL extracts stood out for their content of salvigenin, apigenin-7-O-glucoside, and cirsimaritin, and the *Lavandula* spp. extracts for their content of luteolin-7-O-glucoside, apigenin-7-O-glucoside, and luteolin-7-O-rutinoside. In addition, as previously mentioned, kaempferol was the only flavonol found in TM, and it was not detected in the other solid residues of the MAPs analyzed.

The RO extracts were highlighted for their high phenolic terpene content, accounting for 84% of the total amount of phenolic compounds, with carnosic acid being the most abundant, followed by carnosol, carnosic acid derivative, and 12-methylcarnosic acid. Carnosic acid is a labdane-type diterpene with a strong antioxidant activity, but it is also very unstable in the presence of oxygen [33]. As a result, other phenolic terpenes are formed from the degradation of carnosic acid, including carnosol, methyl carnosate, rosmanol, and rosmadial [29]. Kontogianni et al. [24] attributed the antioxidant activity of fresh RO extracts to the presence of carnosol, rosmarinic acid, and carnosic acid. In addition to the RO extracts, only the SL extracts showed a phenolic terpene content of 46% of the total phenols, but quantitatively, it was 10 times lower than the RO extracts.

A principal component analysis (PCA) was performed to investigate whether the information given by the composition in the phenolic compounds could allow the solid residues obtained from the distillation of the MAPs to be differentiated. The PCA selected four components that explained 75.1% of the total variance (Table S1).

Figure 3a shows the distribution of the samples in the plane defined by the first two dimensions that explained 62.5% of the total variance. As can be observed, all the differences described above in the phenolic profile allowed us to distinguish the solid residues obtained from the distillation of the studied plant species. The RO extracts, cited for the positive side of PC 1 and PC 2, were clearly separated from the other extracts and were positively correlated with the phenolic terpenes; some flavones (mainly glucuronide derivatives of luteolin and apigenin, hispidulin-7-O-glucoside and nepetin-7-O-glucoside); and cryptochlorogenic acid and rosmarinic acid-3-O-glucoside (Table S1). On the other hand, the TM extracts were located on the upper left side of the plane, correlating mainly with kaempferol and flavanones, and some flavones such as aglicones luteolin and apigenin, 6-hydroxyluteolin, 6-hydroxyapigenin, and their O-glycoside derivatives. Finally, the samples belonging to species of the genus *Lavandula* were grouped together and closer to the *S. lavandulifolia* samples, indicating that these two species are closer in terms of chemotaxonomy. These samples were located on the lower side of the plane due to the lowest content of most of the phenolic compounds, with the exception of the phenolic acids, danshensu, salvianolic acid A, coumaric acid-O-glucoside, dihydro-p-coumaric acid glucoside, and two ferulic acid-O-glucosides, and the flavone, luteolin-7-O-rutinoside (Table S1).

Another PCA was performed only with the *Lavandula* spp. samples, and it selected three components that explained 78.3% of the total variance (Table S2). Figure 3b shows the distribution of the *Lavandula* spp. samples in the plane defined by the first two dimensions that explain 70.9% of the total variance. Among the *Lavandula* spp. samples, spike lavender (LL), sited on the right side of the plane, was separated from lavender and lavandin, which were grouped together. The LL extracts were positively correlated with the phenolic acids, danshensu, coumaric acid-O-glucoside, dihydro-p-coumaric acid glucoside, rosmarinic acid, and salvianolic acid A, and the flavones, luteolin-7-O-glucoside, and apigenin-7-O-glucoside (Table S2). Although the concentrations of each phenolic compound can be affected by the population origin or the environmental conditions in which the plants grow, the phenolic profile of the species is preserved. Therefore, chemotaxonomy has been widely used in the *Lamiaceae* family in different genera and has even been used to differentiate taxa at the subspecies level [39,40].

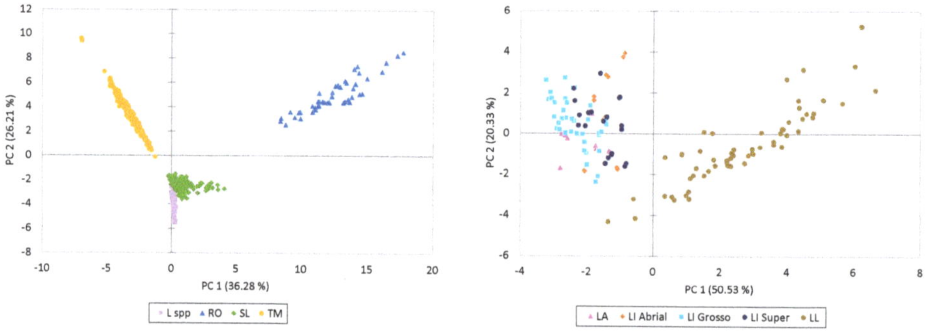

(**a**) (**b**)

Figure 3. Principal component analysis. (**a**) Representation of the samples of solid residues of *Thymus mastichina* L. (TM), *Salvia lavandulifolia* Vahl (SL), *Lavandula* spp. (L. spp.), and *Rosmarinus officinalis* L. (RO); (**b**) representation of the samples of solid residues of *Lavandula latifolia* Medik (LL), three types of *L. x intermedia* Emeric ex Loisel. (LI_ Abrial, LI_Grosso and LI_Super), and *L. angustifolia* Mill. (LA).

3. Materials and Methods

3.1. Plant Samples

Six different aromatic plants were studied: *Thymus mastichina* L. (TM), *Salvia lavandulifolia* Vahl (SL), and *Lavandula latifolia* Medik (LL) and three types of *L. x intermedia* Emeric ex Loisel, including (LI_ Abrial, LI_Grosso and LI_Super), *L. angustifolia* Mill. (LA), and *Rosmarinus officinalis* L. (RO). All the samples analyzed came from cultivation trials located in the experimental farm of ITACyL in Valladolid. The plant material consisted of several wild populations of TM, SL, LL, and RO (8 to 15 different populations depending on the species) and was previously collected with vegetation multiplied around the Iberian Peninsula. LA, LI_ Abrial, LI_Grosso, and LI_Super plant materials were supplied by a commercial nursery. Samples were collected in different assays from 2013 to 2021. A total of 576 samples from different species and years were analyzed.

The aerial parts of wild and/or cultivated populations of *Lamiaceae* species were collected during the blossom phase in different years. The plant material was dried at room temperature in the dark for four weeks after collection. When the drying process was finished, leaves and flowers were separated from the stems, and only the mixture of leaves and flowers was used for distillation. The plant material after the distillation process was used for analysis.

3.2. Chemicals

HPLC-grade reagents and analytical-grade reagents were provided by Panreac (Madrid, Spain). The standard phenolic compounds were purchased from Sigma–Aldrich (Steinheim, Germany), Fluka (Buchs, Switzerland), Phytolab (Vestenbergsgreuth, Germany), Extrasynthèse (Lyon, France), and Targetmol (Wellesley Hills, MA, USA), as has been indicated in Table 2. P-coumaric acid was used as the internal standard (IS) and was provided by Sigma–Aldrich (Steinheim, Germany).

3.3. Solid Residue from the Hydro-Distillation of the Aromatic Plants and Extraction of the Phenolic Compounds

In total, 150 g of each dry sample was hydro-distilled (2 L) for 2.5 h using a Clevenger-type apparatus to extract the EOs. The solid residue obtained after the distillation was dried in an oven at 30 °C ± 2 °C for about 48 h. The dry solid residue was ground using a Retsch ZM 1 centrifugal mill (Fisher Scientific S.L., Madrid, Spain) and sieved between 800 and 280 μm.

In total, 0.25 g of the dry solid residue was extracted four times with 10 mL of methanol at room temperature by shaking in an orbital shaker KS-15 A (Bühler, Switzerland) at 150 rpm for 30 min. In the first extraction, 400 μL of p-coumaric acid was added as an internal standard (IS, 400 mg/L). The four extracts were pooled and filtered through nylon membrane filters at 0.45 μm, 47 mm (Filter-Lab, Barcelona, Spain). The extracts of each sample were concentrated under vacuum at 35 °C using a rotary evaporator R-210 (Büchi, Flawil, Switzerland) to a volume of less than 5 mL. This concentrated extract was transferred to a 5 mL volumetric flask and made up to the mark with methanol. The extracts were filtered (PVDF filters, 0.45 μm) and injected into the chromatograph on the day of extraction.

3.4. High-Performance Liquid Chromatography Diode Array Detection (HPLC-DAD) for the Quantification of Phenolic Compounds

The filtered extracts obtained were analyzed with an Agilent Technologies LC series 1200 (Boblingen, Germany) equipped with a quaternary gradient pump, a thermostated column compartment, an automatic injector, and a Diode Array Detection (DAD) system. The chromatographic separation was achieved on a reverse-phase ACE C18 column (250 mm × 4.6 mm i.d., 5 μm particle size) provided by Symta (Madrid, Spain) and thermostated at 25 °C. The mobile phase was solvent A (20 mM formic acid buffer pH = 2.8) and solvent B (methanol). The gradient was linear at a flow rate of 0.8 mL/min: 0 min, 10% B; 5 min, 20% B; 8 min, 38% B; 18 min, 38% B; 20 min, 70% B; 37 min, 90% B; and 38 min, 100% B and then held for 7 min before returning to the initial conditions.

The injection volume of samples and standards was 100 μL, using an injection program mixing 80 μL of solvent A and the 20 μL sample. The UV spectra were recorded in the range of 200–400 nm range, and chromatograms were recorded at 254, 280, 320, and 350 nm.

The quantification was carried out at different wavelengths depending on each compound (Table 1), using the internal standard method. The compounds tentatively identified where commercial standard was available and were quantified using the calibration curve of the available standard with the most similar UV absorption spectrum belonging to the same phenolic class: chlorogenic acid for peak 3, trans-cinnamic acid for peaks 4 and 8, ferulic acid for peaks 6 and 10, rosmarinic acid for peak 11, luteolin for peaks 9, 20, and 26, apigenin for peaks 22, 24, 35, and 43, and rosmanol for peaks 40 and 42 and carnosic acid for peaks 44, 45, and 47. The phenolic compound contents were expressed in micrograms per gram of dry solid residue weight.

3.5. Liquid Chromatography Diode Array Detection and Mass Spectrometer (LC-DAD-MS) for the Identification of Phenolic Compounds

The identification of the phenolic compounds was carried out on an Agilent LC series 1200 (Boblingen, Germany) equipped with a binary pump, a thermostated column compartment, an automatic injector, a 1260 DAD system, and a Mass Spectrometer Series 6410 Triple Quad. The chromatographic conditions were the same as for HPLC-DAD analyses, except that the flow rate was reduced to 0.6 mL/min. An electrospray ionization (ESI) source was used, and the ESI conditions were as follows: capillary voltage, 4 kV; nebulizer pressure, 45 psi; drying gas flow, 8 L/min; and drying gas temperature, 350 °C. Full mass spectra were recorded in both positive and negative ionization modes in the range of m/z 100–1000, using a cone voltage of 150 V.

The compounds without standards were identified tentatively by comparing the UV spectra characteristics and the mass spectra (molecular weight (m/z) and ion fragmentation) with data available in the literature. Most of the compounds were clearly recognized from the negative ESI-MS spectra, but in some cases, they were identified in the positive mode.

3.6. Method Validation

Calibration, precision, accuracy, limits of detection, and quantification were evaluated for method validation.

The internal standard calibration curves were made using the corresponding phenolic standards at twelve different concentrations, and the linearity and range were determined for each phenolic compound.

The precision was calculated by performing three different extractions of each sample on the same day (repeatability, %RSD, n = 3) and on three different days (reproducibility, %RSD, n = 9), and injecting on each day of the extraction.

The accuracy was calculated by spiking two different samples with known amounts of some standards at two different concentrations and analyzing these samples twice a day on three different days for each added concentration to determine the recovery percentage (% RSD, n = 6). Therefore, the matrix effect was considered.

The limit of detection (LOD) is the lowest concentration at which the instrument can detect but not quantify, and the limit of quantification (LOQ) is the lowest concentration at which the instrument can detect and quantify each compound. These limits were calculated from the standard deviation of the y-interceptions of the regression lines (Sy) and the slope of the calibration curve (S) at levels approaching the LOD ($LOD = 3.3 \times Sy/S$ and $LOQ = 10 \times Sy/S$).

3.7. Statistical Analyses

The concentrations of each phenolic compound were expressed as the mean ± standard deviation (SD) in the solid residues from each MAP. A Box and Whisker Plot were used to represent the data distribution between the solid residues from different Spanish species of *Lamiaceae*.

Principal component analysis (PCA) was carried out to determine similarities or differences between the concentration of the phenolic compounds evaluated in the solid residue of the MAPs studied.

All the statistical analyses were performed using the XLSTAT software (Addinsoft, version 2022.1).

4. Conclusions

A simple, sensitive, reproducible, and accurate HPLC-DAD-MS method is proposed that allows the simultaneous identification and quantification of forty-seven phenolic compounds grouped into phenolic acids, flavonoids (flavones, flavanones, and flavonols), and phenolic terpenes in the solid residue obtained after the EO extraction from different important Spanish species of *Lamiaceae*. The *Rosmarinus officinalis* extracts showed the highest total phenolic content due to the high concentrations of phenolic terpenes. The *Thymus mastichina* extracts were characterized by kaempferol and flavanones, and some flavones derived from luteolin and apigenin. Finally, the samples belonging to species of the genus *Lavandula* were closer to the *Salvia lavandulifolia* extracts, presenting the lowest content of most of the phenolic compounds, with the exception of some phenolic acids, such as danshensu, salvianolic acid A, and glucosides of hydroxycinnamic acids, and the flavone, luteolin-7-O-rutinoside. These results indicate that these solid residues still contain important amounts of different polyphenols, which are antioxidants and can, therefore, be used in different industries.

Therefore, this work provides information on the quantification of a large number of phenolic compounds using a simple methodology and the characterization and valorization of the solid residues obtained from the distillation process of EOs.

Supplementary Materials: The following supporting information can be downloaded at https://www.mdpi.com/article/10.3390/molecules29225285/s1. Table S1: Factor loadings after varimax rotation of all the samples of solid residues obtained from the distillation of the MAPs; Table S2: Factor loadings after varimax rotation of the Lavandula spp. samples of solid residues obtained from the distillation.

Author Contributions: Conceptualization, S.P.-M. and M.C.A.-S.-M.; methodology, S.P.-M., M.B.-H. and M.C.A.-S.-M.; validation, S.P.-M. and M.B.-H.; formal analysis, S.P.-M.; investigation, M.B.-H.; resources, M.C.A.-S.-M.; writing—original draft preparation, S.P.-M.; writing—review and editing, S.P.-M. and M.C.A.-S.-M.; visualization, S.P.-M.; supervision, S.P.-M.; project administration, M.C.A.-S.-M.; funding acquisition, S.P.-M. and M.C.A.-S.-M. All authors have read and agreed to the published version of the manuscript.

Funding: This research was funded by the Instituto Nacional de Investigación y Tecnología Agraria y Alimentaria (INIA, Spain), grant number RTA2017-00031-C43.

Institutional Review Board Statement: Not applicable.

Informed Consent Statement: Not applicable.

Data Availability Statement: Data are contained within the article and Supplementary Materials.

Conflicts of Interest: The authors declare no conflicts of interest.

eferences

1. Brinckmann, J.A.; Kathe, W.; Berkhoudt, K.; Harter, D.E.V.; Schippmann, U. A new global estimation of medicinal and aromatic plant species in commercial cultivation and their conservation status. *Econ. Bot.* **2022**, *76*, 319–333. [CrossRef]
2. Taghouti, I.; Cristobal, R.; Brenko, A.; Stara, K.; Markos, N.; Chapelet, B.; Hamrouni, L.; Bursic, D.; Bonet, J.A. The market evolution of medicinal and aromatic plants: A global supply chain analysis and an application of the Delphi method in the Mediterranean area. *Forests* **2022**, *13*, 808. [CrossRef]
3. Sharma, A.; Gumber, K.; Gohain, A.; Bhatia, T.; Sohal, H.S.; Mutreja, V.; Bhardwaj, G. Chapter 3—Importance of essential oils and current trends in use of essential oils (aroma therapy, agrofood, and medicinal usage). In *Essential Oils. Extraction, Characterization and Applications*; Nayik, G.A., Ansari, M.J., Eds.; Academic Press: Cambridge, MA, USA, 2023; pp. 53–83. [CrossRef]
4. Shabankareh, H.G.; Khorasaninejad, S.; Soltanloo, H.; Shariati, V. Physiological response and secondary metabolites of three lavender genotypes under water deficit. *Sci. Rep.* **2021**, *11*, 19164. [CrossRef] [PubMed]
5. Stashenko, E.; Martínez, J.R. Essential oils and the circular bioeconomy. In *Essential Oils—Recent Advances, New Perspectives and Applications*; Viskelis, J., Ed.; IntechOpen: London, UK, 2024; pp. 1–13. [CrossRef]
6. Saha, A.; Basak, B.B. Scope of value addition and utilization of residual biomass from medicinal and aromatic plants. *Ind. Crops Prod.* **2020**, *145*, 111979. [CrossRef]
7. Galanakis, C.M. *Food Waste Recovery: Processing Technologies and Industrial Techniques*, 1st ed.; Academic Press-Elsevier Inc.: London, UK, 2015.
8. Torras-Claveria, L.; Jauregui, O.; Bastida, J. Antioxidant activity and phenolic composition of lavandin (*Lavandula x intermedia* Emeric ex Loiseleur) waste. *J. Agric. Food Chem.* **2007**, *55*, 8436–8443. [CrossRef]
9. Navarrete, A.; Herrero, M.; Martín, A.; Cocero, M.J.; Ibáñez, E. Valorization of solid wastes from essential oil industry. *J. Food Eng.* **2011**, *104*, 196–201. [CrossRef]
10. Parejo, I.; Viladomat, F.; Bastida, J.; Rosas-Romero, A.; Flerlage, N.; Burillo, J.; Codina, C. Comparison between the radical scavenging activity and antioxidant activity of six distilled and nondistilled mediterranean herbs and aromatic plants. *J. Agric. Food Chem.* **2002**, *50*, 6882–6890. [CrossRef]
11. Asensio-S.-Manzanera, M.C.; Martín, H.; Sanz, M.A.; Herrero, B. Antioxidant activity of *Lavandula latifolia*, *Salvia lavandulifolia* and *Thymus mastichina* collected in Spain. *Acta Hortic.* **2011**, *925*, 281–290. [CrossRef]
12. Santana-Méridas, O.; González-Coloma, A.; Sánchez-Vioque, R. Agricultural residues as a source of bioactive natural products. *Phytochem. Rev.* **2012**, *11*, 447–466. [CrossRef]
13. Moisa, C.; Copolovici, L.; Bungau, S.; Pop, G.; Imbrea, I.; Lupitu, A.; Nemeth, S.; Copolovici, D. Wastes resulting from aromatic plants distillation. Biosources of antioxidants and phenolic compounds with biological active principles. *Farmacia* **2018**, *66*, 289–295.
14. Ortiz de Elguea-Culebras, G.; Melero Bravo, E.; Sánchez-Vioque, R. Potential sources and methodologies for the recovery of phenolic compounds from distillation residues of Mediterranean aromatic plants. An approach to the valuation of by-products of the essential oil market—A review. *Ind. Crops Prod.* **2022**, *175*, 114261. [CrossRef]
15. Miljanović, A.; Dent, M.; Grbin, D.; Pedisić, S.; Zorić, Z.; Marijanović, Z.; Jerković, I.; Bielen, A. Sage, Rosemary, and Bay Laurel Hydrodistillation By-Products as a Source of Bioactive Compounds. *Plants* **2023**, *12*, 2394. [CrossRef] [PubMed]
16. Mena, P.; Cirlini, M.; Tassotti, M.; Herrlinger, K.A.; Dall'Asta, C.; Del Rio, D. Phytochemical profiling of flavonoids, phenolic acids, terpenoids, and volatile fraction of a rosemary (*Rosmarinus officinalis* L.) extract. *Molecules* **2016**, *21*, 1576. [CrossRef] [PubMed]

17. Contreras, M.M.; Algieri, F.; Rodríguez-Nogales, A.; Gálvez, J.; Segura-Carretero, A. Phytochemical profiling of anti-inflammato Lavandula extracts via RP–HPLC–DAD–QTOF–MS and –MS/MS: Assessment of their qualitative and quantitative difference *Electrophoresis* **2018**, *39*, 1284–1293. [CrossRef]
18. Irakli, M.; Skendi, A.; Bouloumpasi, E.; Chatzopoulou, P.; Biliaderis, C.G. LC-MS Identification and quantification of phenol compounds in solid residues from the essential oil industry. *Antioxidants* **2021**, *10*, 2016. [CrossRef]
19. Almela, L.; Sánchez-Muñoz, B.; Fernández-López, J.A.; Roca, M.J.; Rabea, V. Liquid chromatograpic–mass spectrometric analys of phenolics and free radical scavenging activity of rosemary extract from different raw material. *J. Chromatogr. A* **2006**, *112* 221–229. [CrossRef]
20. Clifford, M.N.; Wu, W.; Kirkpatrick, J.; Kuhnert, N. Profiling the chlorogenic acids and other caffeic acid derivatives of herba chrysanthemum by LC-MSn. *J. Agric. Food Chem.* **2007**, *55*, 929–936. [CrossRef]
21. Borrás-Linares, I.; Stojanović, Z.; Quirantes-Piné, R.; Arráez-Román, D.; Švarc-Gajić, J.; Fernández-Gutiérrez, A.; Segura-Carretero, A *Rosmarinus officinalis* leaves as a natural source of bioactive compounds. *Int. J. Mol. Sci.* **2014**, *15*, 20585–20606. [CrossRef]
22. Sharma, Y.; Velamuri, R.; Fagan, J.; Schaefer, J. Full-Spectrum analysis of bioactive compounds in Rosemary (*Rosmarinus officinali* L.) as influenced by different extraction methods. *Molecules* **2020**, *25*, 4599. [CrossRef]
23. Achour, M.; Mateos, R.; Fredj, M.B.; Mtiraoui, A.; Bravo, L.; Saguem, S. A comprehensive characterisation of Rosemary te obtained from *Rosmarinus officinalis* L. Collected in a sub-humid area of Tunisia. *Phytochem. Anal.* **2018**, *29*, 87–100. [CrossRef]
24. Kontogianni, V.G.; Tomic, G.; Nikolic, I.; Nerantzaki, A.A.; Sayyad, N.; Stosic-Grujicic, S.; Stojanovic, I.; Gerothanassis, I.P Tzakos, A.G. Phytochemical profile of *Rosmarinus officinalis* and *Salvia officinalis* extracts and correlation to their antioxidant anc anti-proliferative activity. *Food Chem.* **2013**, *136*, 120–129. [CrossRef] [PubMed]
25. Grayer, R.J.; Eckert, M.R.; Lever, A.; Veitch, N.C.; Kite, G.C.; Paton, A.J. Distribution of exudate flavonoids in the genus Plectranthus. *Biochem. Syst. Ecol.* **2010**, *38*, 335–341. [CrossRef]
26. Psarrou, I.; Oreopoulou, A.; Tsimogiannis, D.; Oreopoulou, V. Extraction kinetics of phenolic antioxidants from the hydrc distillation residues of rosemary and effect of pretreatment and extraction parameters. *Molecules* **2020**, *25*, 4520. [CrossRef] [PubMed]
27. Cuvelier, M.E.; Richard, H.; Berset, C. Antioxidative activity and phenolic composition of pilot-plant and commercial extracts of sage and rosemary. *J. Am. Oil Chem.* **1996**, *73*, 645–652. [CrossRef]
28. Schwarz, K.; Ternes, W. Antioxidative constituents of *Rosmarinus officinalis* and *Salvia officinalis*. II. Isolation of carnosic acid and formation of other phenolic diterpenes. *Z. Lebensm. Unters. Forsch.* **1992**, *195*, 99–103. [CrossRef]
29. Doolaege, E.H.A.; Raes, K.; Smet, K.; Andjelkovic, M.; Van Poucke, C.; De Smet, S.; Verhe, R. Characterization of two unknown compounds in methanol extracts of rosemary oil. *J. Agric. Food Chem.* **2007**, *55*, 7283–7287. [CrossRef]
30. Zhang, Y.; Smuts, J.P.; Dodbiba, E.; Rangarajan, R.; Lang, J.C.; Armstrong, D.W. Degradation study of carnosic acid, carnosol, rosmarinic acid, and rosemary extract (*Rosmarinus officinalis* L.) assessed using HPLC. *J. Agric. Food Chem.* **2012**, *60*, 9305–9314. [CrossRef]
31. Bicchi, C.; Binello, A.; Rubiolo, P. Determination of phenolic diterpene antioxidants in rosemary (*Rosmarinus officinalis* L.) with different methods of extraction and analysis. *Phytochem. Anal.* **2000**, *11*, 236–242. [CrossRef]
32. Troncoso, N.; Sierra, H.; Carvajal, L.; Delpiano, P.; Gunther, G. Fast high-performance liquid chromatography and ultraviolet-visible quantification of principal phenolic antioxidants in fresh rosemary. *J. Chromatogr. A* **2005**, *1100*, 20–25. [CrossRef]
33. Borrás-Linares, I.; Arráez-Román, D.; Herrero, M.; Ibáñez, E.; Segura-Carretero, A.; Fernández-Gutiérrez, A. Comparison of different extraction procedures for the comprehensive characterization of bioactive phenolic compounds in *Rosmarinus officinalis* by reversed-phase high-performance liquid chromatography with diode array detection coupled to electrospray time-of-flight mass spectrometry. *J. Chromatogr. A* **2011**, *1218*, 7682–7690. [CrossRef]
34. Pérez-Magariño, S.; Ortega-Heras, M.; Cano-Mozo, E. Optimization of a solid-phase extraction method using copolymer sorbents for isolation of phenolic compounds in red wines and quantification by HPLC. *J. Agric. Food Chem.* **2008**, *56*, 11560–11570. [CrossRef] [PubMed]
35. Petersen, M.; Simmonds, M.S.J. Rosmarinic acid. *Phytochemistry* **2003**, *62*, 121–125. [CrossRef] [PubMed]
36. Hui, X.; Yan-Li, L.; Hong-Cui, T.; Wen-Ting, Z. Preparation and chemical structure identification of methylated metabolites of salvianolic acid A. *Chin. J. Anal. Chem.* **2014**, *42*, 65–70.
37. Li, X.; Yu, C.; Lu, Y.; Gu, Y.; Lu, J.; Xu, W.; Xuan, L.; Wang, Y. Pharmacokinetics, tissue distribution, metabolism, and excretion of depside salts from *Salvia miltiorrhiza* in rats. *Drug Metab. Dispos.* **2007**, *35*, 234–239. [CrossRef]
38. Ma, L.; Tang, L.; Yi, Q. Salvianolic Acids: Potential Source of Natural Drugs for the Treatment of Fibrosis Disease and Cancer. *Front Pharmacol.* **2019**, *10*, 97. [CrossRef]
39. Koshovyi, O.; Raal, A.; Kovaleva, A.; Myha, M.; Ilina, T.; Borodina, N.; Komissarenko, A. The phytochemical and chemotaxonomic study of *Salvia* spp. growing in Ukraine. *J. Appl. Biol. Biotech.* **2020**, *8*, 29–36. [CrossRef]
40. Staszek, D.; Orłowska, M.; Waksmundzka-Hajnos, M.; Sajewicz, M.; Kowalska, T. Marker fingerprints originating from TLC and HPLC for selected plants from the *Lamiaceae* family. *J. Liq. Chromatogr. R. T.* **2013**, *36*, 2463–2475. [CrossRef]

molecules

MDPI

ticle

otential of Sorghum Seeds in Alleviating Hyperglycemia,)xidative Stress, and Glycation Damage

ora Ben El Mahdi [1], Laurent Lemée [2], Quentin Blancart Remaury [2], Lilian Eloy [2], Naima Nhiri [3], aoufal Lakhssassi [4], Francesco Cacciola [5,*] and Mohamed Nhiri [1,*]

1 Laboratory of Biochemistry and Molecular Genetics, Faculty of Sciences and Technologies of Tangier, BP 416, Tangier 90000, Morocco
2 IC2MP-UMR CNRS 7285, Institut de Chimie des Milieux et Matériaux de Poitiers, University of Poitiers, 86073 Poitiers CEDEX, France; laurent.lemee@univ-poitiers.fr (L.L.); quentin.blancart.remaury@univ-poitiers.fr (Q.B.R.); lilian.eloy@univ-poitiers.fr (L.E.)
3 Institute for the Chemistry of Natural Substances, CNRS, Paris Saclay University, 91190 Gif-sur-Yvette, France; naima.nhiri@cnrs.fr
4 School of Agricultural Sciences, College of Agricultural, Life and Physical Sciences, Southern Illinois University, Carbondale, IL 62901, USA; naoufal.lakhssassi@siu.edu
5 Department of Biomedical, Dental, Morphological and Functional Imaging Sciences, University of Messina, 98125 Messina, Italy
* Correspondence: cacciolaf@unime.it (F.C.); mnhiri@uae.ac.ma (M.N.)

Abstract: Diabetes mellitus, characterized by dysregulated glucose metabolism, oxidative stress, and the formation of advanced glycation end products, poses a significant global health burden. In this study, we explored the potential of sorghum (*Sorghum bicolor*) seeds, known for their abundant phytochemical composition, as a natural remedy for diabetes and its associated damage. High-performance liquid chromatography/high-resolution mass spectrometry analysis revealed a remarkable phenolic richness in sorghum grains, including gallic acid, quercetin, and the predominant procyanidin B-1, with ecotype-specific variations in flavonoid distribution. Elemental analysis by ICP showed an abundance of macro-elements (Ca, K, Mg), trace elements (Fe, Mn, Si, Zn), and ultra-trace elements (B, Co, Cr, Cu, Mo, Se, V) essential for human health, supporting its therapeutic and nutritional potential. Additionally, the results demonstrated variable total phenolic contents (188–297 mg GAE/g dE) and total flavonoid contents (66–78 mg QE/g dE), with corresponding differences in antioxidant activities across the five ecotypes. Treatment with sorghum seed extract (SE1) significantly reduced oxidative stress markers, such as malondialdehyde (MDA)by 40% and hydrogen peroxide (H_2O_2) by 63%, in diabetic mice, compared to untreated diabetic controls. Moreover, sorghum extracts exhibited a remarkable increase in antioxidant enzyme activities, including a 50% increase in superoxide dismutase (SOD) activity and a 60% increase in glutathione peroxidase (GPx) activity, indicating their potential to bolster antioxidant defenses against diabetes-induced oxidative stress. These findings underscore the therapeutic potential of sorghum seeds in diabetes management and prevention, paving the way for the development of functional foods with enhanced health benefits.

Keywords: anti-glycation; reactive oxygen species; phytochemical composition; antioxidant enzymes; sorghum bicolor ecotypes seeds

1. Introduction

Diabetes mellitus, characterized by chronic hyperglycemia due to deficiencies in insulin secretion, action, or both, poses a significant global health concern, with 463 million adults affected in 2019 and projections indicating a rise to 700 million by 2045 [1]. Glycation, a non-enzymatic reaction between reducing sugars and proteins, lipids, or nucleic acids, leads to the formation of advanced glycation end products (AGEs), which contribute to diabetic complications such as nephropathy, retinopathy, neuropathy, and cardiovascular diseases [2].

Natural products, particularly plant-derived phenolic compounds, have shown promise in mitigating glycation and AGE formation while reducing the side effects of pharmaceutical drugs due to their antioxidant and anti-inflammatory properties [3,4]. Sorghum (*Sorghum bicolor*), a widely cultivated cereal crop, is rich in phenolic compounds and has demonstrated significant antioxidant, anti-inflammatory, and antidiabetic effects, making it a promising candidate for diabetes management [5]. Recent studies have highlighted the health-promoting properties of sorghum seeds [6,7].

This study aims to assess the therapeutic potential of ethanolic extracts from five different ecotypes of sorghum seeds in diabetes management, focusing on their ability to mitigate glycation and AGE formation. We will examine the phytochemical composition of the seed extracts, their antioxidant activity, and their anti-glycation properties both in vitro and in vivo using an alloxan-induced murine model of diabetes. By elucidating the bioactivity mechanisms of sorghum seed extracts, particularly their inhibition of glycation and AGE formation, this research aims to support the use of natural products in diabetes management and contribute to the development of novel dietary supplements or pharmaceutical agents for diabetes prevention and treatment.

2. Results and Discussion

2.1. Phytochemical Profile and Characterization of Bioactive Compounds Using HPLC-HRMS

Phenolic compounds are widely recognized for their health benefits, including antioxidant, anti-inflammatory, and anticancer effects. Sorghum (*Sorghum bicolor*) seeds, a staple cereal crop, are rich in phenolics and offer potential health benefits. Although several studies have examined the phenolic composition of sorghum seeds, comprehensive profiling using advanced techniques like high-performance liquid chromatography coupled with high-resolution mass spectrometry (HPLC-HRMS) is needed. This study aims to characterize the phenolic compounds in sorghum seed ethanolic extract using HPLC-HRMS and compare our findings with the existing literature.

HPLC-HRMS analysis (Table 1/Figure 1) revealed a rich array of phenolic compounds in sorghum seed ethanolic extract, consistent with previous studies. Identified phenolic acids include gallic acid, vanillic acid, protocatechuic acid, hydroxybenzoic acid, caffeic acid, and ferulic acid, demonstrating the diverse chemical composition of sorghum seeds. Flavonoids identified include procyanidin B-1, catechin, eriodictyol, taxifolin, naringenin, apigenin, quercetin, and rutin, enriching the phenolic profile of sorghum seeds. Our findings align closely with recent reports on sorghum phenolics using HPLC-HRMS and reinforce the robustness of our analytical approach [8–11]. Ecotype-specific variations observed are consistent with those reports, highlighting the importance of genetic and environmental factors in sorghum phenolic research [10,11].

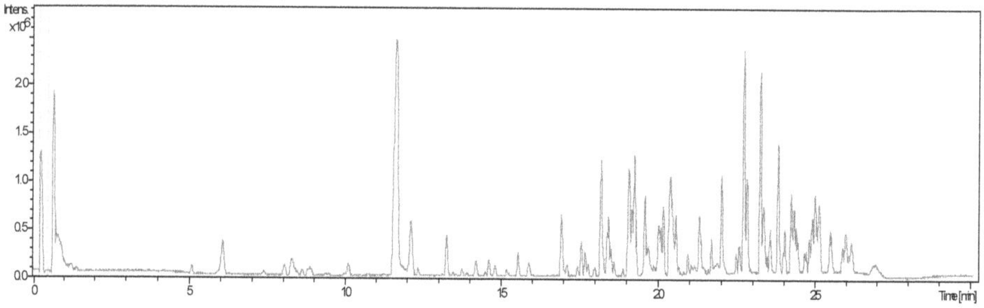

Figure 1. Chromatogram (HPLC-HRMS) of flavonoid compounds in seed ethanolic extract of sorghum ecotype 1.

Table 1. Characteristics of the compounds identified by HPLC-HRMS in the five ecotypes of sorghum seed ethanolic extract.

					Ecotypes				
eak #	t_R (min)	$[M+H]^+$	Name	Family	E1 (%)	E2 (%)	E3 (%)	E4 (%)	E5 (%)
1	0.7	171.0288	Gallic acid	I	0.5 ± 0.1 [a]	0.5 ± 0.1 [a]	0.4 ± 0.1 [a]	-	-
2	1.1	169.0454	Vanillic acid	I	0.9 ± 0.2 [a]	0.3 ± 0.1 [b]	-	-	-
3	1.5	165.0546	p-coumaric acid	II	1.1 ± 0.2 [a]	1.5 ± 0.2 [a]	0.9 ± 0.1 a	1.0 ± 0.1 [a]	-
4	3.8	155.03388	Protocatechic acid	I	1.7 ± 0.3 [a]	2.9 ± 0.4 [b]	1.0 ± 0.2 [a]	1.0 ± 0.1 [a]	0.8 ± 0.1 [a]
5	5.1	579.1497	Procyanidin B-1	V	11.2 ± 1.1 [a]	8.4 ± 0.9 [b]	10.3 ± 1.0 [a]	3.2 ± 0.3 [c]	11.5 ± 1.1 [a]
6	6.1	139.03897	Hydroxybenzoic acid	I	-	-	-	-	1.2 ± 0.2
7	6.1	291.0863	Catechin	V	53.8 ± 5.4 [a]	40.8 ± 4.1 [b]	54.6 ± 5.5 [a]	19.6 ± 2.0 [c]	68.6 ± 6.8 [d]
8	7.0	181.0495	Caffeic acid	II	0.9 ± 0.1 [a]	0.4 ± 0.1 [b]	0.9 ± 0.1 [a]	0.8 ± 0.1 [a]	1.3 ± 0.1 [a]
9	7.1	195.06518	Ferulic acid	II	1.3 ± 0.2 [a]	1.3 ± 0.2 [a]	0.6 ± 0.1 [b]	0.5 ± 0.1 [b]	0.7 ± 0.1 [b]
10	8.2	289.07066	Eriodictyol	IV	5.6 ± 0.6 [a]	6.7 ± 0.9 [b]	2.4 ± 0.3 [b]	1.0 ± 0.1 [c]	-
11	9.1	305.06557	Taxifolin	V	6.8 ± 0.7 [a]	9.7 ± 1.1 [a]	7.9 ± 0.8 [a]	38.4 ± 3.9 [b]	3.3 ± 0.3 [c]
12	9.4	273.07575	Naringenin	IV	1.9 ± 0.2 [a]	2.4 ± 0.2 [a]	0.8 ± 0.1 [b]	0.5 ± 0.1 [b]	-
13	9.5	271.0601	Apigenin	VI	2.3 ± 0.3 [a]	4.3 ± 0.4 [b]	4.5 ± 0.5 [b]	5.7 ± 0.6 [b]	1.9 ± 0.2 [a]
14	10.6	303.04993	Quercetin	VI	0.2 ± 0.1 [a]	2.3 ± 0.3 [a]	1.9 ± 0.2 [a]	9.5 ± 1.0 [b]	0.7 ± 0.1 [b]
15	10.6	611.16066	Rutin	IV	1.6 ± 0.2 [a]	2.3 ± 0.3 [b]	1.1 ± 0.1 [a]	-	1.7 ± 0.2 [ab]

Letters after the values indicate significantly different groups according to the statistical analysis ANOVA followed by Tukey HSD post hoc test. Values are expressed as mean ± SD (standard deviation); I: Hydroxybenzoic acids; II: Hydroxycinnamic acids; IV: Flavanones and Flavanols; V: Dihydroflavonols; VI: Flavones and Flavan-4-ol.

The analysis of flavonoid composition across five sorghum ecotypes (Figure 2) reveals a diverse array of flavonoids. Procyanidin B-1, a condensed tannin with remarkable antioxidant properties, is the predominant compound. Variations in flavonoid distribution among ecotypes were observed: flavanols dominate in ecotypes 1 and 3, flavonols in ecotypes 2 and 5, and flavanonols in ecotype 4. These profiles underscore the interplay between genetic diversity and environmental factors in flavonoid biosynthesis. Other noteworthy flavonoids identified include catechin, quercetin, and kaempferol, all known for their health-promoting properties. Comparative analysis with the existing literature reaffirms the presence of these flavonoids in sorghum seeds, underscoring their potential antioxidant, anti-inflammatory, and anticancer benefits [11].

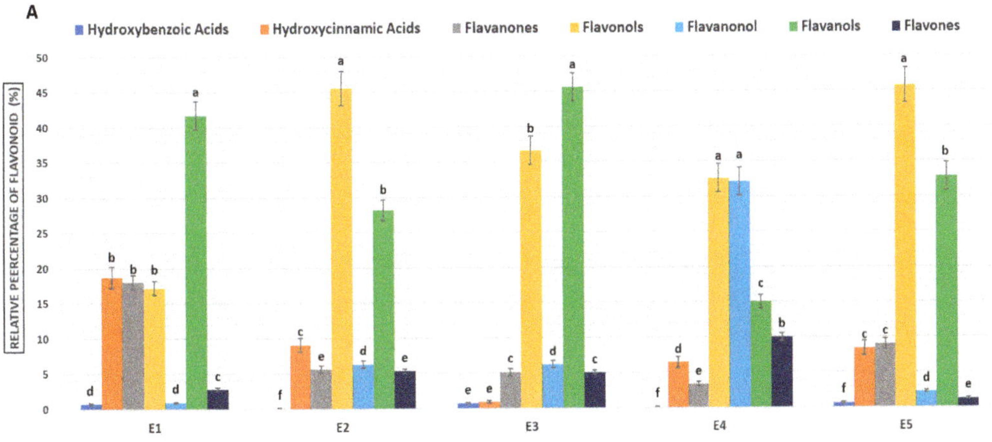

Figure 2. *Cont.*

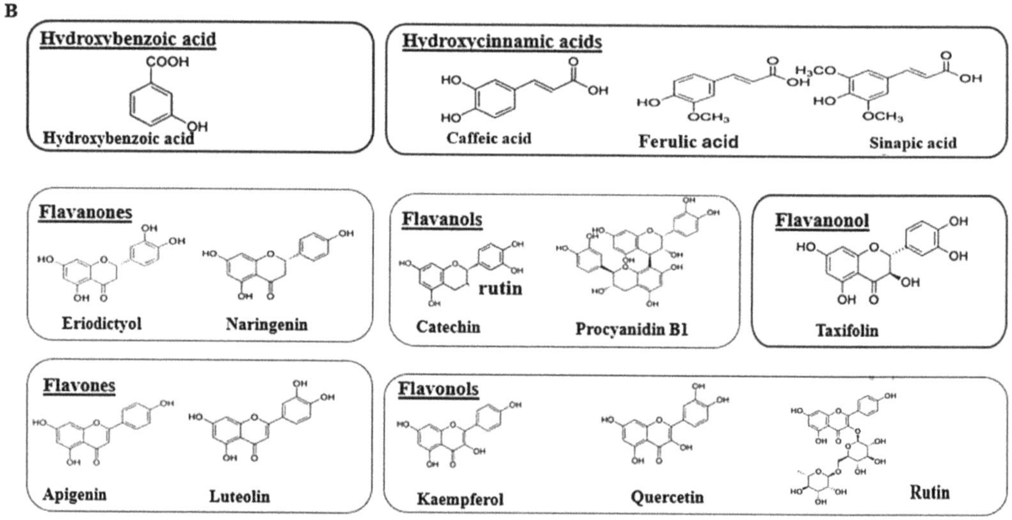

Figure 2. Compounds Identified by HPLC-HRMS in 5 sorghum seed ecotypes. (**A**) Relative percentage of identified compounds in the five sorghum ecotypes (E1, E2, E3, E4, E5), including hydroxybenzoic acids (I), hydroxycinnamic acids (II), flavanones (III), flavonols (IV), flavanonols (V), flavan-4-ols (VI), and flavones (VII). Letters above the bars indicate significant differences between compound classes for each ecotype separately according to ANOVA followed by Tukey's HSD post hoc test, with significance letters assigned in descending order of means. (**B**) Chemical structures of the identified compound families.

2.2. Elemental Analysis of Mineral Composition of Sorghum by ICP

Elemental analysis of sorghum seeds provides a detailed portrait of their mineral composition, offering valuable insights into the potential therapeutic benefits of this cereal crop. Across different ecotypes, significant variations in macroelements, microelements, and ultra-trace elements underscore the rich diversity of bioactive compounds in sorghum plants.

Macroelements, such as calcium (Ca), potassium (K), and magnesium (Mg), exhibit notable concentrations, with ecotype 1 particularly standing out with Ca levels at 199 mg/kg, K at 510 mg/kg, and Mg at 260 mg/kg (Table 2). These essential minerals are integral to fundamental physiological processes, including bone density, muscle contraction, and nerve signaling. The substantial abundance of macroelements in sorghum seeds positions this cereal as a natural reservoir of vital nutrients essential for overall health and vitality, consistent with recent findings [12,13].

Table 2. Major and trace elements determined by inductively coupled plasma coupled with optical emission spectroscopy (ICP-OES).

Ecotype	Macroelements (mg/kg)					Microelements (mg/kg)			
	Ca	K	Mg	P	S	Fe	Mn	Si	Zn
E1	199 ± 0.5	510 ± 2.0	260 ± 3.0	530 ± 2.5	100 ± 1.6	57 ± 0.3	28 ± 0.2	396 ± 5.9	33 ± 0.6
E2	78 ± 0.5	300 ± 6.0	100 ± 0.9	220 ± 0.9	100 ± 1.2	30 ± 0.4	13 ± 0.1	408 ± 1.7	14 ± 0.1
E3	64 ± 0.8	240 ± 4.0	80 ± 0.7	180 ± 1.4	100 ± 0.9	18 ± 0.2	8.7 ± 0.02	375 ± 5.2	12 ± 0.3
E4	104 ± 0.4	250 ± 3.0	120 ± 0.9	250 ± 2.3	120 ± 1.3	33 ± 0.2	16 ± 0.1	472 ± 12.1	20 ± 0.3
E5	125 ± 0.8	530 ± 4.2	170 ± 5.0	420 ± 1.4	100 ± 0.5	195 ± 2.2	22 ± 0.1	361 ± 5.7	23 ± 0.9

Microelement variability further enriches sorghum's therapeutic potential, with iron (Fe) concentrations reaching up to 57 mg/kg and manganese (Mn) up to 28 mg/kg. Silicon (Si) content spans from 361 to 472 mg/kg, while zinc (Zn) levels range from 12 to 33 mg/kg (Table 2). Iron and manganese are pivotal for oxygen transport and antioxidant defense mechanisms, respectively, while silicon and zinc contribute significantly to bone integrity and immune function. These microelements in sorghum seeds offer promising avenues for supporting cardiovascular health, bolstering immune resilience, and managing oxidative stress, as highlighted in previous studies [14,15].

The discovery of ultra-trace elements further underscores sorghum's nutritional complexity and therapeutic promise. Boron (B) concentrations range from 1.09 to 2.15 mg/kg, cobalt (Co) is found in trace amounts (<100 μg/kg), chromium (Cr) levels span from 68 to 688 μg/kg, copper (Cu) ranges from 1.36 to 4.41 mg/kg, molybdenum (Mo) from 0.38 to 0.96 mg/kg, selenium (Se) from 12 to 23 μg /kg, and vanadium (V) from 16 to 397 μg/kg (Table 3). These ultra-trace elements are associated with various health benefits, including antioxidative properties, improved metabolic regulation, and modulation of thyroid function. Incorporating sorghum-based foods into the diet offers a holistic approach to harnessing these essential nutrients, thereby promoting overall well-being and vitality, as supported by recent investigations [16,17].

Table 3. Ultra-trace elements determined by inductively coupled plasma coupled with mass spectrometry (ICP-MS).

Ecotype	^{11}B (mg/Kg)	^{111}Cd (μg/kg)	^{59}Co (μg/kg)	^{52}Cr (μg/kg)	^{63}Cu (mg/Kg)	^{95}Mo (mg/Kg)	^{23}Na (mg/Kg)	^{60}Ni (mg/Kg)	^{208}Pb (μg/kg)	^{78}Se (μg/kg)	^{51}V (μg/kg)
E1	1.62 ± 0.03	<2	33 ± 1.38	204 ± 3.87	4.41 ± 0.04	0.52 ± 0.02	9.80 ± 0.22	0.45 ± 0.01	121 ± 1.69	12 ± 1.44	70 ± 1.58
E2	1.09 ± 0.04	<2	16 ± 1.04	192 ± 3.07	1.91 ± 0.04	0.380 ± 0.003	3.28 ± 0.22	0.340 ± 0.005	30 ± 0.72	15 ± 1.44	41 ± 2.04
E3	1.58 ± 0.05	<2	12 ± 0.84	68 ± 1.97	2.22 ± 0.02	0.70 ± 0.01	4.83 ± 0.60	0.156 ± 0.002	29 ± 0.66	16 ± 1.71	16 ± 1.38
E4	2.15 ± 0.03	<2	24 ± 0.98	249 ± 2.49	2.65 ± 0.05	0.96 ± 0.03	2.98 ± 0.10	0.56 ± 0.02	30 ± 0.63	23 ± 2.71	43 ± 2.92
E5	1.48 ± 0.04	<2	98 ± 2.54	688 ± 10.32	1.36 ± 0.01	0.61 ± 0.01	2.12 ± 0.02	0.67 ± 0.01	207 ± 3.72	17 ± 1.34	397 ± 10.95

While the present study focused on the beneficial properties of sorghum seed extracts, such as their antioxidant and antidiabetic activities, it is crucial to consider the safety of these extracts for practical applications. ICP analyses revealed the presence of essential and trace elements, with the absence of potentially toxic elements at concerning levels, such as cadmium (Cd) and lead (Pb) (Tables 2 and 3). However, while these results are reassuring, they are insufficient to assess the extracts' overall toxicity. To ensure the safety of sorghum extracts, comprehensive toxicological studies should be conducted. These studies should include:

Evaluation of Acute and Chronic Toxicity: Acute and chronic toxicity studies on animal models to determine safe dosage ranges and identify potential short-term and long-term adverse effects.

Cytotoxicity Tests: In vitro cytotoxicity assays using various cell lines to assess potential toxic effects at the cellular level.

Monitoring of Side Effects: Monitoring potential side effects, including allergic reactions, gastrointestinal disorders, and liver or kidney toxicity.

Finally, although the data presented in Table 2 were obtained from single measurements without replicates, the observed values are consistent with the existing literature and remain reliable. Future studies should include multiple biological and technical replicates to ensure statistical robustness and validate these findings.

2.3. Total Phenolic and Flavonoid Content

The total phenolic and flavonoid content of sorghum's ecotype extracts, as presented in Table 4, reveals insights into the phytochemical composition of sorghum seeds. The variations observed in polyphenols, flavonoids, and tannins across different ecotypes highlight the diversity of bioactive compounds present in sorghum plants.

Table 4. Polyphenol, flavonoid, and tannin content of sorghum ecotype extracts.

Ecotypes	Extract Yield (%)	Polyphenols (TPC) (mg GAE/g dE)	Flavonoids (TFC) (mg QE/g dE)	Tannins EAT (mg QE/mg dE)
E1	26.73 [b]	297 ± 0.008 [b]	72 ± 0.002 [a]	0.253 ± 0.002 [a]
E2	17.05 [a]	188 ± 0.006 [a]	66 ± 0.005 [a]	0.210 ± 0.009 [b]
E3	22.49 [a]	229 ± 0.006 [a]	78 ± 0.002 [a]	0.203 ± 0.002 [c]
E4	25.23 [a]	262 ± 0.005 [a]	73 ± 0.001 [a]	0.222 ± 0.003 [d]
E5	19.12 [a]	192 ± 0.005 [a]	70 ± 0.001 [a]	0.195 ± 0.001 [e]

Values are mean ± standard deviation. Statistical groupings are based on Tukey's HSD test. Values followed by the same letter within a column are not significantly different ($p > 0.05$). GAE: Gallic Acid Equivalent. QE: Quercetin Equivalent. dE: dry extract.

When we compare our findings with the existing literature, we observe consistency in certain trends. For instance, ecotype 1 exhibited the highest extract yield, consistent with prior studies [18,19]. Regarding polyphenol content, ecotype 1 showed the highest concentration, while ecotype 2 had the lowest, aligning with previous reports [20,21].

Likewise, discernible differences in flavonoid content were observed among the various ecotypes, with ecotype 3 exhibiting the highest concentration and ecotype 2 demonstrating the lowest. This consistency with previous studies [22,23] suggests a pattern influenced by genetic and environmental factors.

Furthermore, tannin content varied among ecotypes, with ecotype 1 exhibiting the highest levels, followed by ecotype 4. Recent findings [22,23] support these trends, highlighting the multifaceted influence of genetic and environmental factors on sorghum's phytochemical profile.

In conclusion, the analysis of total phenolic and flavonoid content underscores the need to consider genetic and environmental factors in determining the phytochemical composition of sorghum seeds. Further research should focus on elucidating the biosynthesis mechanisms of these bioactive compounds and their potential health benefits.

2.4. Antioxidant Activities

The antioxidant properties of sorghum ecotypes, as presented in Table 5, reveal their potential health benefits. Among the ecotypes, E4 exhibited the strongest DPPH scavenging activity, with the lowest IC_{50} value of 0.043 mg/mL, indicating robust antioxidant potential. Conversely, E2 showed the highest IC_{50} value of 0.06 mg/mL, suggesting relatively weaker antioxidant activity than the others. These findings align with previous research highlighting variability in sorghum antioxidant activity [11,22].

Table 5. Antioxidant activities of sorghum ecotype ethanolic extracts.

Ecotypes	Antioxidant Properties (IC_{50} Values; mg/mL)			
	DPPH Scavenging Activity	ABTS	Metal Chelating Activity	Reducing Power (mg AAE/g dE)
E1	0.059 ± 0.002 [a]	0.090 ± 0.010 [a]	2.57 ± 0.10 [a]	111 ± 8 [a]
E2	0.060 ± 0.001 [a]	0.131 ± 0.003 [b]	1.816 ± 0.07 [b]	121 ± 8 [b]
E3	0.059 ± 0.002 [a]	0.118 ± 0.002 [b]	1.3 ± 0.1 [b]	119 ± 3 [b]
E4	0.043 ± 0.004 [b]	0.118 ± 0.020 [b]	2.110 ± 0.007 [a]	101 ± 7 [c]
E5	0.048 ± 0.001 [b]	0.140 ± 0.010 [b]	1.297 ± 0.020 [b]	90 ± 3 [c]
Trolox	0.203 ± 0.020 [c]	0.086 ± 0.001 [c]	-	-
EDTA	-	-	0.189 ± 0.030 [c]	-

All values are mean ± standard deviation. Different letters indicate significant differences (ANOVA, Tukey's test, $p < 0.05$) between conditions in the same test. IC_{50}: the extract concentration providing 50% inhibition; DPPH: 2.2-diphenyl-1-picrylhydrazyl; ABTS: 2.2′-azino-bis (3-ethylbenzothiazoline-6-sulphonic acid); AAE: Ascorbic acid Equivalent; dE: dry extract.

In addition to DPPH scavenging, ABTS assay results demonstrated E1's potent activity of 0.090 mg/mL, contrasting with E5's weaker activity of 0.140 mg/mL, consistent with prior studies [23].

Metal chelating activity, crucial for preventing oxidation reactions, varied among ecotypes. E5 exhibited the highest activity of 1.297 mg/mL, while E1 showed the lowest activity of 2.57 mg/mL, aligning with previous findings [22].

Reducing power, indicating electron donation ability, ranged from 90.16 mg AAE/g dE to 121.813 mg AAE/g dE. E5 displayed the highest and E2 the lowest, consistent with diverse antioxidant potentials.

Comparatively, Trolox, a well-known antioxidant standard, showed DPPH scavenging activity of 0.203 mg/mL, indicating that some sorghum ecotypes, like E4, possess competitive antioxidant potential. EDTA, used as a standard, showed activity of 0.189 mg/mL for metal chelating activity, suggesting that some sorghum ecotypes also have significant metal chelating capabilities.

The identified bioactive compounds (Figure 1, Table 1) can neutralize free radicals through electron transfer and proton donation reactions, reducing oxidative stress. Specific reactivity tests and metabolic pathway analyses can provide further insights into these mechanisms. Additionally, utilizing cellular models to observe the effects of sorghum extracts at the cellular level would help understand the changes in oxidative stress markers and the cellular signaling pathways involved in the antioxidant response

The diverse antioxidant capacities observed among sorghum ecotypes might significantly affect combating diabetes-related complications such as advanced glycation end products (AGEs) and reactive oxygen species (ROS) production. These antioxidant properties may help mitigate oxidative stress, a key factor in the progression of diabetes and its associated health issues. By scavenging free radicals and inhibiting ROS-mediated damage, sorghum antioxidants hold promise in alleviating diabetic complications, including cardiovascular diseases, nephropathy, and neuropathy. Sorghum antioxidants mitigate oxidative stress by enhancing the activity of key antioxidant enzymes such as superoxide dismutase (SOD), catalase, and glutathione peroxidase (GPx). These enzymes are crucial in neutralizing reactive oxygen species (ROS) and protecting cellular components from oxidative damage. Further research is needed to explore the specific mechanisms underlying how sorghum antioxidants interact with pathways related to AGE formation and ROS generation. Understanding these mechanisms will enable the development of targeted interventions using sorghum-based products to improve health outcomes in individuals with diabetes.

2.5. *In Vitro Antidiabetic Activity: α-Amylase and α-Glucosidase Inhibition Effects*

Sorghum-E extracts' in vitro antidiabetic activity from five ecotypes was assessed using α-amylase and α-glucosidase inhibition assays. These enzymes are crucial for carbohydrate digestion and glucose absorption, making them prime targets for antidiabetic treatments. Sorghum extracts may reduce hyperglycemia by inhibiting the activity of carbohydrate-hydrolyzing enzymes such as α-amylase and α-glucosidase. By slowing down the breakdown of carbohydrates into glucose, sorghum extracts can help manage postprandial blood glucose levels, thereby acting as a natural approach to glycemic control. The inhibitory effects were quantified by determining IC_{50} values, representing the concentration of extract required to inhibit 50% of enzyme activity (Table 6).

Table 6. Inhibition results of sorghum-E extract on both α-amylase and α-glucosidase enzymes.

Ecotypes	IC50 (mg/mL)	
	Alpha-Amylase	**Alpha-Glucosidase**
Sorghum-E1	1.02 ± 0.20 [a]	2.013 ± 0.008 [a]
Sorghum-E2	1.54 ± 0.03 [a]	1.07 ± 0.06 [a]
Sorghum-E3	2.87 ± 0.01 [a]	1.038 ± 0.040 [a]
Sorghum-E4	3.4 ± 0.1 [a]	1.889 ± 0.020 [a]
Sorghum-E5	2.92 ± 0.06 [a]	2.009 ± 0.020 [a]
Acarbose	0.05 ± 0.02 [b]	0.399 ± 0.003 [b]

Values are means ± standard deviation. Different letters in the same column indicate significant differences ($p < 0.05$) according to Tukey's multiple range test. For example, values followed by the same letter are not significantly different, while those followed by different letters are significantly different. Comparing the IC^{50} values of sorghum-E extracts to acarbose, a known α-glucosidase inhibitor used as a positive control, revealed that sorghum-E extracts from all ecotypes exhibited inhibitory activity against both enzymes, albeit with varying potency. For instance, sorghum-E1 and sorghum-E2 extracts displayed IC_{50} values of 1.021 mg/mL and 1.536 mg/mL, respectively, against α-amylase, indicating potent inhibition. Similarly, sorghum-E2, sorghum-E3, and sorghum-E5 extracts exhibited IC_{50} values ranging from 1.038 mg/mL to 1.889 mg/mL against α-glucosidase, demonstrating significant inhibitory activity.

The potent inhibitory activity of sorghum seed extracts against α-amylase and α-glucosidase enzymes suggests their promising role as a natural inhibitor for controlling postprandial blood glucose levels. This aligns with findings from recent studies investigating the antidiabetic properties of sorghum extracts [24–27]. These observations imply that sorghum seed extracts could potentially regulate carbohydrate digestion and absorption, contributing to glycemic control. Therefore, further exploration of sorghum-derived compounds as potential therapeutic agents for glycemic management and related metabolic conditions is warranted.

2.6. *In Vitro Antiglycation Activities*

Figure 3A illustrates the percentage of inhibition of advanced glycation end products (AGEs) by different sorghum ecotypes across varying concentrations, alongside a positive control (Metformin). AGEs, formed through non-enzymatic glycation, contribute to numerous pathophysiological processes, notably diabetes and cardiovascular diseases. The bioactive compounds in sorghum, including flavonoids and phenolic acids, can inhibit the formation of advanced glycation end products (AGEs) by interacting with the glycation pathways. These compounds can bind to free sugars or amino groups in proteins, preventing the non-enzymatic reactions that lead to AGE formation. Additionally, they can act through several other mechanisms, such as neutralizing free radicals and reducing oxidative stress, which is a key driver in AGE formation. Sorghum bioactives can also neutralize reactive intermediates like methylglyoxal, a precursor in AGE formation, and chelate metal ions such as iron and copper, which catalyze the formation of AGEs.

Examining the inhibitory effects across different concentrations underscores dose-dependent responses. At lower concentrations (0.16 and 0.31 mg/mL), the sorghum ecotypes demonstrated modest inhibitory effects, with ecotypes 3 and 5 showing slightly higher inhibition. As concentrations increased, so did inhibitory activity against AGE formation. Ecotype 5 consistently exhibited the highest inhibition at higher concentrations (2.50 and 5.00 mg/mL). Compared with Metformin, sorghum extracts displayed comparable or superior inhibitory activity against AGE formation, suggesting their potential as anti-glycation agents. Variability in inhibitory activity among sorghum ecotypes may stem from differences in phytochemical composition. Sorghum contains various bioactive compounds, including phenolic acids and flavonoids, known for their antioxidant and anti-glycation properties. These differences in composition likely influence inhibitory activity against AGE formation. These findings highlight sorghum's potential as a functional food with anti-glycation properties. Further research is needed to identify specific bioactive compounds responsible for inhibition and explore their mechanisms of action in mitigating AGE-related complications. Clinical studies are warranted to validate sorghum's therapeutic potential in managing diabetes and other AGE-associated disorders.

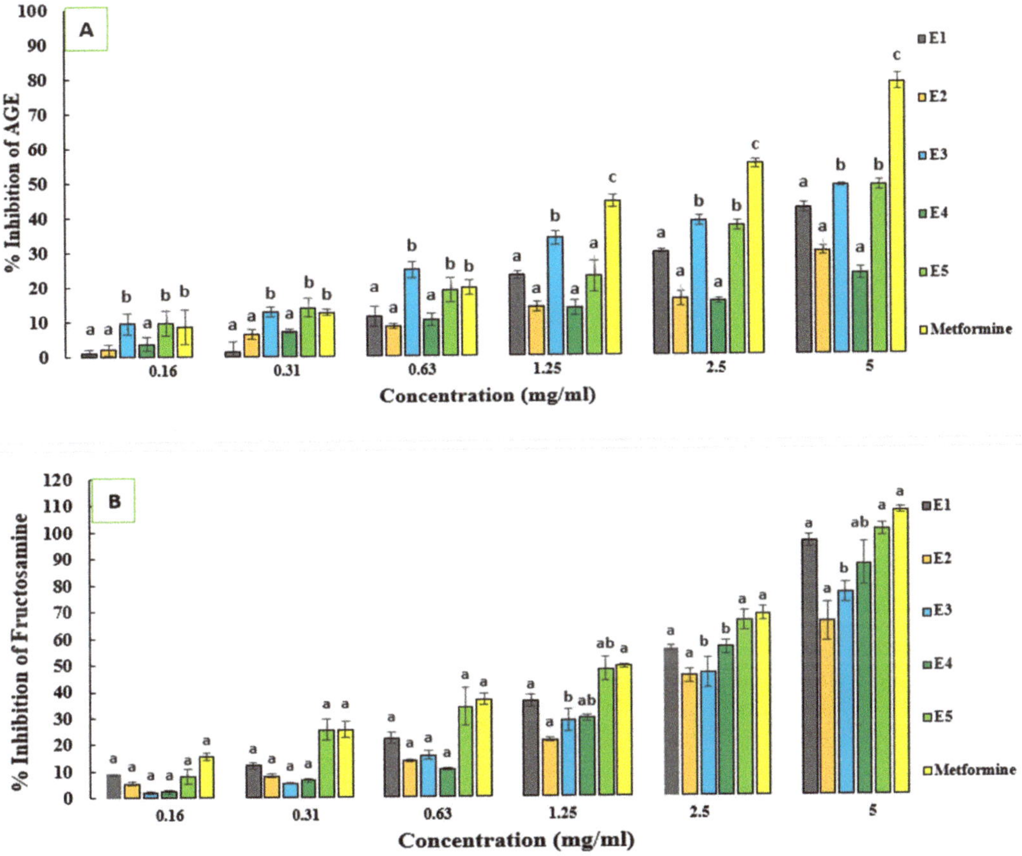

Figure 3. (**A**) Percentage of inhibition of AGE formation and (**B**) percentage of inhibition of fructosamine in the BSA/fructose system. Each value represents the mean of three replicates. Bars represent the standard error. Different letters indicate a significant difference between conditions of the same concentration at $p < 0.05$.

Moreover, the IC_{50} values reported by Yilmaz et al. [28] for sorghum extracts (ranging from 13.93 to 63.00 µg/mL) are notably lower than the concentrations observed in our study. This indicates potentially higher potency of the sorghum extracts tested by Yilmaz et al. in inhibiting AGE formation than the sorghum ecotypes examined in our study.

Farrar et al. [29] also demonstrated that ethanolic extracts of certain sorghum bran varieties containing anthocyanins, luteolinidin, and apigeninidin significantly inhibited protein glycation. While our study did not specifically investigate these compounds, the presence of phenolic compounds in sorghum, including flavonoids, aligns with previous findings suggesting their role in inhibiting protein glycation.

In conclusion, sorghum emerges as a promising candidate in food chemistry, offering potent anti-glycation properties that merit further exploration. Understanding the mechanisms underlying sorghum's effects and its specific bioactive compounds is essential for leveraging its potential in preventing chronic diseases.

Figure 3B presents the concentration of sorghum extract (ecotypes 1–5) and Metformin, along with their respective percentage inhibition on fructosamine levels. Fructosamine is a marker of glycemic control, reflecting average blood glucose.

Across all ecotypes (1–5), as the concentration of sorghum extract increases, there is a notable increase in the percentage inhibition of fructosamine levels. This indicates a dose-dependent response, suggesting that higher concentrations of sorghum extracts lead to more significant inhibition of fructosamine, indicative of improved glycemic control. Among the different ecotypes, there are variations in the efficacy of fructosamine inhibition. For instance, ecotype five consistently exhibits the highest percentage of inhibition across all concentrations compared to other ecotypes.

This suggests that ecotype 5 may contain bioactive compounds with stronger anti-glycemic properties than other ecotypes. Ecotypes 1 and 3 also demonstrate significant inhibition of fructosamine, albeit to a lesser extent compared to ecotype 5. However, ecotypes 2 and 4 show comparatively lower inhibition percentages, indicating potentially lower concentrations of bioactive compounds associated with glycemic control.

The efficacy of sorghum extracts for glycemic control, particularly at higher concentrations, warrants comparison to Metformin, a widely used anti-diabetic medication [30,31]. Metformin remains the first-line therapy for type 2 diabetes, showing superior efficacy in lowering blood glucose levels and reducing complications. Nevertheless, concerns about Metformin's long-term safety prompt the exploration of alternative therapies like sorghum extracts [32,33]. While sorghum extracts demonstrate promising anti-glycemic effects, especially at higher concentrations, Metformin consistently exhibits a high percentage inhibition of fructosamine across all concentrations, often comparable to or exceeding the inhibition observed with sorghum extracts. Further investigation is necessary to compare the bioavailability, safety profiles, and long-term effects of sorghum extracts and Metformin on glycemic control.

Our results, dose-dependent inhibition of fructosamine levels, are consistent with previous research demonstrating their anti-diabetic properties due to their rich bioactive compound content [34,35]. Different sorghum ecotypes may vary in efficacy, potentially due to variations in phytochemical profiles, necessitating further research for standardization [33,35].

2.7. In Vivo Antidiabetic Effect of Sorghum Seeds in Alloxan-Induced Diabetic Mice

2.7.1. Effect of Sorghum Seed Extract on Fasting Blood Glucose Level and Body Weight in Diabetic Mice

The experiment involved inducing severe hyperglycemia in mice through alloxan injection, resulting in initial blood glucose levels of 501 mg/dL for the diabetic non-treated group (Diabetic.NT) at week 0 (Table 7). Subsequent treatment with sorghum seed extract at doses of 125 mg/kg and 250 mg/kg led to a gradual decrease in blood glucose levels over three weeks, with levels dropping to 287.89 mg/dL and 165.54 mg/dL, respectively, by week 3. Notably, both doses of sorghum seed extract significantly reduced blood glucose levels compared to the diabetic non-treated group. Furthermore, the higher dose (+250 mg/kg) demonstrated a more substantial reduction in blood glucose levels compared to the lower dose (+125 mg/kg), indicating a dose-dependent response. Additionally, the standard antidiabetic drug glibenclamide exhibited a significant but comparatively lower reduction in blood glucose levels by week 3. These results suggest that sorghum seed extract, particularly at higher doses, effectively lowers blood glucose levels in alloxan-induced diabetic mice, potentially rivaling or surpassing the efficacy of glibenclamide. Further investigations comparing sorghum seed extract with other antidiabetic agents are warranted to fully elucidate its therapeutic promise and safety profile.

Table 7. Effect of sorghum seed extract ecotype 1 on blood glucose level and body weight of alloxan-induced diabetic mice.

Treatment	Blood Glucose Level (mg/dL) Time (By Weeks)				Body Weight (g) Time (By Weeks)			
	Week 0	Week 1	Week 2	Week 3	Week 0	Week 1	Week 2	Week 3
Normal	78.01 ± 3.2 ^{c}	70.00 ± 3.03 ^{e}	77.65 ± 1.2 ^{c}	73.43 ± 5 ^{d}	21.12 ± 1.87 ^{b}	25.02 ± 4.02 ^{b}	22.91 ± 4.3 ^{b}	23.12 ± 0.23 ^{b}
iabetic.NT	501 ± 9.00 ^{a}	508.02 ± 32.0 ^{a}	467.71 ± 43.9 ^{a}	421.76 ± 26.7 ^{a}	19.23 ± 3.01 ^{a}	17.32 ± 2.3 ^{a}	18.12 ± 3.05 ^{a}	16.88 ± 3.93 ^{a}
125 mg/Kg	532 ± 54 ^{a}	433 ± 12 ^{b}	389.02 ± 65.0 ^{b}	287.89 ± 42 ^{b}	18.87 ± 3.87 ^{a}	18.23 ± 4.02 ^{a}	17.91 ± 3.4 ^{a}	19.54 ± 1.05 ^{a}
250 mg/Kg	356.9 ± 28 ^{b}	350.7 ± 65.1 ^{c}	277.7 ± 23.7 ^{b}	165.54 ± 9.02 ^{c}	17.01 ± 3.65 ^{a}	17.65 ± 3.22 ^{a}	17.13 ± 4.06 ^{a}	19.66 ± 1.24 ^{a}
libenclamide	444.83 ± 43.02 ^{a}	384.30 ± 54.3 ^{d}	282.02 ± 17.8 ^{b}	199.76 ± 23.1 ^{c}	21.76 ± 4.3 ^{a}	20.22 ± 2.43 ^{a}	19.83 ± 3.90 ^{a}	20.94 ± 3.46 ^{a}

The data are presented as mean ± SEM. Statistical analysis was performed using one-way ANOVA followed by Tukey's post hoc test for multiple comparisons. Different letters indicate statistically significant differences among the treatments ($p < 0.05$).

2.7.2. Effect of Sorghum Seed Extract on Biochemical Parameters in Diabetic Mice

Figure 4 presents the impact of sorghum seed extract (ecotype 1) (SE1) on key biochemical parameters in alloxan-induced diabetic mice, juxtaposed against non-diabetic control mice and untreated diabetic counterparts (Diabetic NT).

The levels of malondialdehyde (MDA) and hydrogen peroxide (H_2O_2) content in different treatment groups are as follows: In the NT diabetic group, elevated levels of MDA and H_2O_2 content are observed compared to the non-diabetic control, indicating increased oxidative stress associated with diabetes.

Treatment with sorghum seed extract at both doses shows a significant reduction in MDA and H_2O_2 content compared to the NT diabetic group. This suggests a potential antioxidative effect of sorghum seed extract in mitigating oxidative stress induced by diabetes. Interestingly, the reduction in MDA and H_2O_2 content appears to be dose-dependent, with the group treated with a higher dose (250 mg/kg) exhibiting a more pronounced decrease in oxidative stress markers compared to the group treated with a lower dose (125 mg/kg). Moreover, the antioxidative effect of sorghum seed extract seems comparable to that of glibenclamide, a standard antidiabetic medication, particularly at the higher dose. This suggests that sorghum seed extract may possess antioxidative properties that could contribute to its potential therapeutic benefits in diabetes management. Sorghum extracts may exert anti-inflammatory effects by modulating signaling pathways such as NF-κB and MAPK, which are involved in the inflammatory response. By reducing inflammation, sorghum extracts can further alleviate oxidative stress and protect against complications associated with chronic inflammation in diabetes.

Comparisons with the existing literature on the antioxidative effects of sorghum seed extract in diabetic animal models could provide further insights into the findings and validate its potential as a natural antioxidant therapy for diabetes-related oxidative stress. For instance, studies by Wu et al. (2011) and Ajiboye et al. (2013) reported similar reductions in MDA and H_2O_2 levels following treatment with sorghum seed extract in diabetic rats, supporting our observations [36,37]. However, further investigations are warranted to elucidate the underlying mechanisms and establish the efficacy of sorghum seed extract as an adjunct therapy for diabetes [38]. The presence of specific bioactive compounds like procyanidin B-1 in sorghum contributes significantly to its antioxidant and antiglycation activities. Procyanidin B-1 is known for its ability to scavenge free radicals and inhibit protein glycation, highlighting its potential therapeutic benefits in managing diabetes and its complications.

Superoxide Dismutase (SOD) Activity: Diabetic mice exhibited diminished SOD activity compared to non-diabetic controls, corroborating previous studies indicating oxidative stress in diabetes. Notably, SE1 seed treatment, particularly at 250 mg/kg, resulted in a significant increase in SOD activity, aligning with previous findings showcasing the antioxidant potential of sorghum extracts in alleviating oxidative stress in diabetic models.

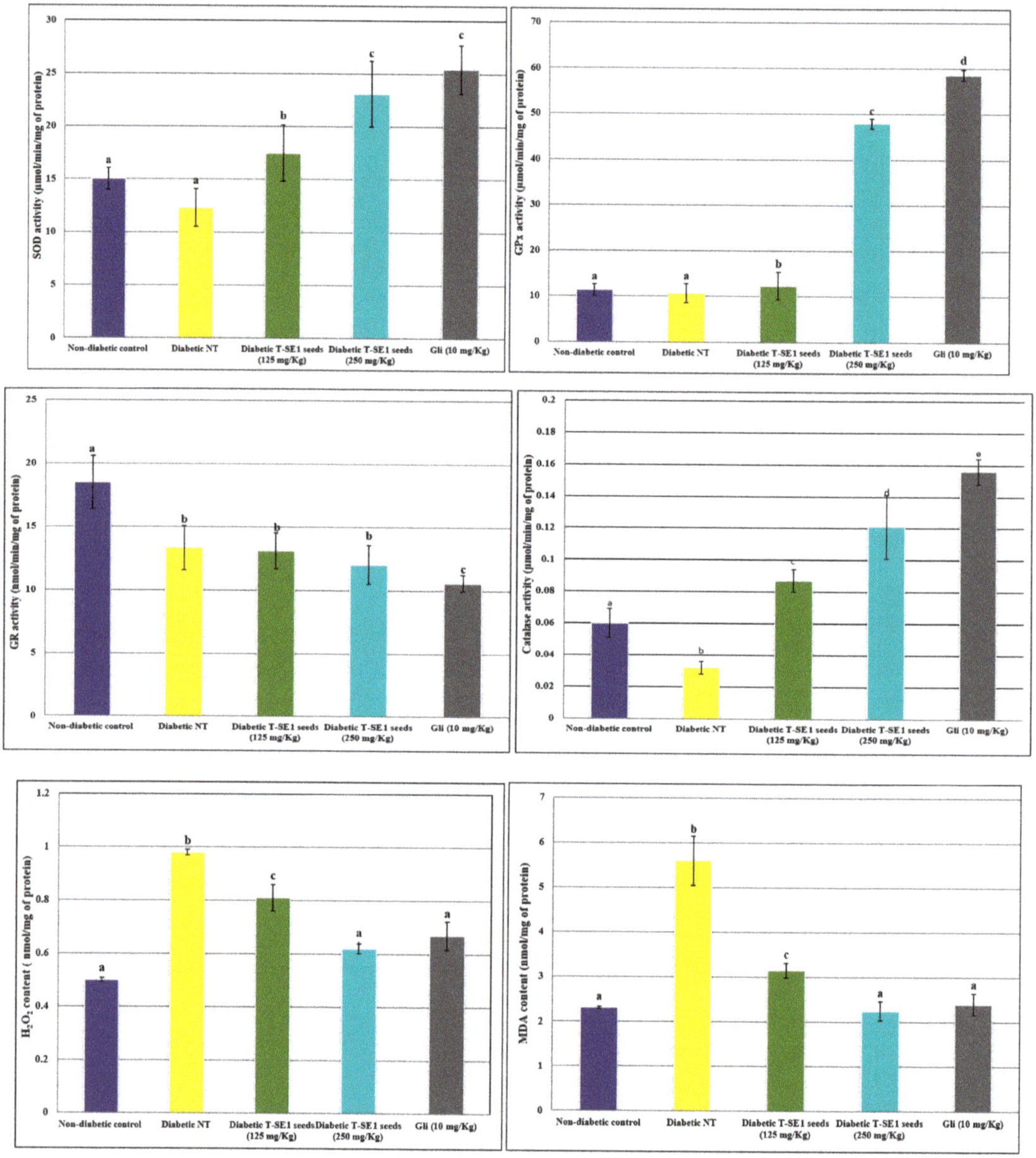

Figure 4. Effect of SE1 seeds on the activities of antioxidant enzymes in the liver of alloxan-induced diabetic mice. Abbreviations are SOD: superoxide dismutase; CAT: catalase; GR: glutathione reductase; GPx: glutathione peroxidase; MDA: malondialdehyde; H_2O_2: hydrogen peroxide. Bars represent the standard error. Different letters indicate significant differences among treatments at $p < 0.05$.

Catalase Activity: Reduced catalase activity observed in diabetic mice underscores impaired antioxidant defenses linked to diabetes-induced oxidative stress, consistent with prior research. SE1 seed treatment exhibited a dose-dependent elevation in catalase activity, akin to recent findings of Xiong et al. [39], suggesting the antioxidative prowess of sorghum extracts in bolstering antioxidant defenses against diabetic-induced oxidative stress.

Glutathione Peroxidase (GPx) Activity: A decline in GPx activity in diabetic mice reflects compromised antioxidant capacity, corroborating studies implicating oxidative stress in diabetes progression. SE1 seed treatment, especially at 250 mg/Kg, significantly elevated GPx activity, mirroring the findings of Li et al. (2020) [40], highlighting the potential dose-dependent influence of sorghum extracts in enhancing antioxidant defenses against diabetic-induced oxidative stress.

Glutathione Reductase (GR) Activity: Reduced GR activity in diabetic mice suggests perturbed redox homeostasis associated with diabetes, consistent with the literature on diabetic complications. While SE1 seed treatment showed potential in mitigating oxidative stress, further research is needed to fully elucidate its effects on GR activity and its contribution to antioxidant defenses in diabetic conditions, aligning with the call for comprehensive investigation in diabetic models [41] by Johnson et al. (2022).

In summary, the outcomes underscore the potential of sorghum seed extract (SE1) in mitigating diabetic-induced oxidative stress through modulation of crucial antioxidant enzyme activities, with findings resonating with the existing literature. However, further research is warranted to elucidate the specific bioactive compounds responsible for these effects and to explore SE1's therapeutic potential in managing diabetic complications.

The results presented highlight the potential of sorghum seed extract as a natural therapeutic agent for managing diabetes and related complications, offering insights into its mechanisms of action and comparative efficacy against conventional treatments. The inhibitory activity of sorghum extracts against α-amylase and α-glucosidase enzymes, essential for carbohydrate digestion and glucose absorption, demonstrates their role in regulating postprandial blood glucose levels. These findings align with the recent literature, affirming sorghum's antidiabetic properties attributed to its rich bioactive compound content.

Furthermore, this study elucidates sorghum's anti-glycation properties, evidenced by its inhibitory effects on AGE formation. The dose-dependent response underscores sorghum's potential as an anti-glycation agent, with certain ecotypes exhibiting superior inhibitory activity. Notably, sorghum extracts display comparable or superior efficacy to conventional medications like Metformin, warranting further investigation into their therapeutic potential and safety profiles.

Additionally, the antioxidative effects of sorghum seed extract offer promising avenues for combating oxidative stress associated with diabetes. The significant reductions in oxidative stress markers and enhancement of crucial antioxidant enzyme activities underscore sorghum's role in bolstering antioxidant defenses. These findings are consistent with previous research, validating sorghum as a natural antioxidant therapy for diabetes-related oxidative stress.

The results of this study indicate that sorghum seed extracts possess significant antioxidant and anti-glycation properties, suggesting their potential as ingredients in functional foods. Incorporating these extracts into food products could provide benefits for managing diabetes and preventing associated complications. For example, sorghum-based products enriched with these extracts could be developed to help regulate blood glucose levels and reduce oxidative stress in diabetic patients. Moreover, the nutritional and therapeutic properties of sorghum seeds could be exploited to create dietary supplements aimed at improving overall metabolic health. These developments would not only diversify food options for individuals with diabetes but also promote healthier eating habits in the general population.

3. Materials and Methods

3.1. Plant Material and Extraction

The plant material consisted of 5 sorghum ecotypes (Figure 5) selected from a collection of 10 sorghum ecotypes [42] by Bouargalne et al. (2022), conserved at Laboratory of Biochemistry and Molecular Genetics, Faculty of Sciences and Technologies, Abdelmalek Essaadi University Tangier (Morocco).

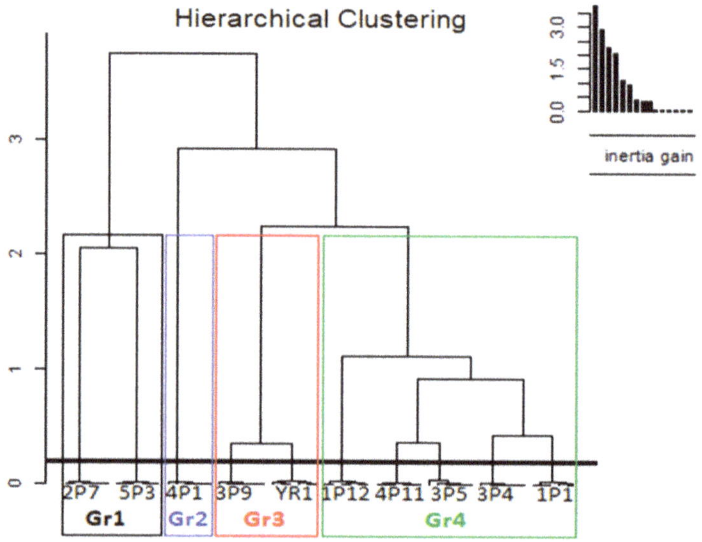

Figure 5. Hierarchical clustering of the 5 Moroccan sorghum seed ecotypes based on agromorphological traits.

This core collection is representative of the Moroccan cultivated sorghum varieties, which were sampled based on geographical origin. The study was conducted in Tinghir (17°25′ N latitude and 78° E longitude and minimum and maximum temperatures of 10 °C and 45 °C with an annual total rainfall of 619 mm) during 2015 summer season. Sowing was performed following a randomized complete block design (RCBD) with three replicates in 8 m rows with a row spacing 60 cm apart and plant-to-plant spaced at 20 cm. Standard cultivation practices followed locally were adopted. NPK fertilizer was applied at the rate of 100 kg/ha. Data on the morpho-agronomic traits were collected from six randomly selected individual plants for each ecotype. Meanwhile, the collected mature seeds were used for further biochemical experiments. The 25 morpho-agronomic traits that were measured or scored are described in Appendix A (Table A1) [42]. The whole seeds were washed with distilled water, shade dried, and powdered with a mechanical grinder; the powder was stored in an airtight container until needed for further use. Five distinct ecotypes (E1, E2, E3, E4, and E5) (Table 8) were subjected to extraction using 80% ethanol. The resulting homogenate underwent a 24 h incubation in darkness under continuous agitation at 260 rpm. Subsequently, the extract was filtered through Whatman filter paper and centrifuged at 5000× *g* for 10 min. Finally, evaporation at 40 °C yielded extracts with the following respective percentages for E1, E2, E3, E4, and E5: 26.73%, 17.05%, 22.49%, 25.23%, and 19.12%.

Table 8. Description of the 5 studied sorghum seed ecotypes.

Abbreviations of the Studied Ecotypes	Extraction Solvent	Designation of Sorghum Ecotype	Groups	Number of Ecotypes Per Group
E1	Ethanol	1P12	Gr4	5
E2	Ethanol	5P3	Gr1	2
E3	Ethanol	3P4	Gr4	5
E4	Ethanol	3P9	Gr3	2
E5	Ethanol	4P1	Gr2	1

To ensure the long-term stability and reusability of the extracts, stability tests were conducted by storing the extracts at 4 °C and at room temperature. The chemical composition and antioxidant activity of the extracts were monitored over a period of six months. Additionally, reusability tests were performed by subjecting the extracts to multiple cycles of use in various applications, such as food preservation and pharmaceutical formulations, to evaluate their practical viability

3.2. *HPLC-HRMS*

HPLC-HRMS analysis was conducted utilizing a Thermo Scientific Ultimate 3000 system (Thermo Fisher Scientific, Waltham, MA, USA) coupled with a Bruker Impact HD Series Quadrupole-Time of Flight (QTOF) (Bruker, Billerica, MA, USA) mass spectrometer equipped with an Electrospray (ESI) ion source. The chromatographic separation was performed on an Acquity UPLC® (Waters Corporation, Milford, MA, USA) BEH 1.7 µm column (100 × 2.1 mm), with a mobile phase flow rate of 0.4 mL min-1 and the column temperature maintained at 30 °C. The mobile phase consisted of ultrapure water (eluent A) and methanol (eluent B) containing 0.1% formic acid. Following the injection of a 10 µL sample, the gradient began with 1% B and remained constant for 2 min. Subsequently, it ramped up to 99% B over 23 min and remained constant for 2 min.

For the mass spectrometry method, the ESI source operated in positive mode with the following parameters: gas temperature of 200 °C, drying gas flow rate of 8.0 L min^{-1}, nebulizer pressure of 2.1 Bar, capillary voltage of 2500 V, and a mass range from 50 to 1000 m/z.

3.3. *Elemental Analysis by ICP*

3.3.1. Preparation Sample

Approximately 300 mg of the sorghum kernel ground to a fine and homogenized powder was weighed into a Teflon vessel. Two digestions were realized for each sample with different acid mixtures, 8.0 mL HNO_3 and 7.0 mL HNO_3 + 1.0 mL HF for silicon analysis were added then the vessel was sealed. An Anton Paar Multiwave (Graz, Austria) pro fitted with a rotor for 8 pressure vessels NXF100 was used. The temperature and pressure limit in the digestion program was fixed at 230 °C and 60 bars. The digestate was made up to a volume of 50 mL using de-ionized water. De-ionized 18 $M\Omega.cm^{-1}$ water used for the preparation of all blanks, standards, and sample solutions was obtained from a Millipore water purification system (Merck Millipore, Burlington, MA, USA). External calibration curves were performed using certified calibration standard solutions and diluted according to the working ranges required.

3.3.2. Major and Trace Elements Using ICP-OES

Quantification of major (K, Mg, P, S) and minor (Ca, Fe, Mn, Si, Zn) elements was performed by inductively coupled plasma optical spectrometry (ICP-OES; Agilent 5110 VDV, Agilent Technologies, Santa Clara, CA, USA). The sample introduction system was constituted of a Seaspray™ nebulizer (Glass Expansion, Port Melbourne, Victoria, Australia) and a quartz double-pass cyclonic chamber. The analyses were conducted at IC2MP-UMR CNRS 7285, Institute of Chemistry of Environments and Materials of Poitiers, University of Poitiers, 86073 Poitiers CEDEX, France For silicon, an inert spray chamber was used to facilitate the analysis of HF digests without prior neutralization and avoid quartz sample introduction attacks from free HF. The parameter settings were as follows: Argon (Ar) plasma gas flow, 12 $L.min^{-1}$; Ar auxiliary gas flow, 0.7 $L.min^{-1}$; Ar nebulizer gas flow, 0.70 $L.min^{-1}$; and radio frequency power 1200 W.

To avoid most spectral interferences, two wavelengths for each element were checked. An internal standard (ISTD) solution containing 1 mg/L of yttrium was delivered online to the sample before nebulization using a Y-connector. ISTD allows us to check and correct non-spectral interference.

3.3.3. Ultra-Trace Elements Using ICP-MS

Quantification of trace elements (11B, 111Cd, 59Co, 52Cr, 63Cu, 95Mo, 23Na, 60Ni, 208Pb, 78Se and 51V) was performed by inductively coupled plasma mass spectrometry (ICP-MS Agilent 7800). Operating conditions are shown in Table 9. Spectrometer parameters were optimized daily using a 1 µg/L Agilent tuning solution. The spectrometer was equipped with a collision reaction cell (CRC) to reduce interferences. Selenium 78 was analyzed with hydrogen gas in CRC, and H_2 flow gas rate was optimized to eliminate 38Ar40Ar interference. An additional interference on Se from rare-earth element doubly charged (156Gd^{2+}) was avoided by adding 156Gd in the analyte list and calculating the applied Gd^{2+}/Gd ratio correction. Lead was measured with the sum of the 3 most abundant isotopes 206, 207, and 208.

Table 9. ICP-MS Agilent 7800 operating conditions.

ICP-MS Agilent 7800 Operating Conditions			
Collision/reaction cell gas	-	He	H_2
Collision/reaction cell gas flow rate mL.min^{-1}	-	4.3	4.2
Energy discrimination (V)	5	3	5
Nebulizer gas flow rate L.min^{-1}		0.99	
Auxiliary gas flow rate L.min^{-1}		0.9	
Plasma gas flow rate L.min^{-1}		15	
Nebulizer type		Concentric Micromist	
Spray chamber		Quartz Scott double pass	
Injector		Quartz 2.5 mm	
RF Power (W)		1550	
Wash (s)		120	
Integration time (s)		1	
Replicates		4	
Sweeps		100	

3.4. Total Phenolic Content

The total phenolic content was determined by the Folin–Ciocalteu (FC) method, following Singleton and Rossi [43] with some modifications [44]. Briefly, 100 µL of the sample solution was mixed with 400 µL of FC reagent and 1 mL of 7% sodium carbonate (Na_2CO_3) solution. The final volume was adjusted to 1.6 mL with distilled water. The reaction mixture was incubated in the dark for 30 min, followed by absorbance measurement at 725 nm using a spectrophotometer (EPOCH, BioTek) against a blank. The TPC of the extracts was expressed as milligrams of gallic acid equivalents per gram of dry weight (mg GAE/g dw) using a calibration curve generated with gallic acid as the standard.

3.5. Total Flavonoid Content

The flavonoid content was assessed following the method of Huang et al. [45] with some modifications. Briefly, 40 µL of each sample was mixed with 10 µL of acetate potassium (1 M) and 10 µL of aluminum chloride (10%). Next, 100 µL of methanol 50% was added and the total volume was made up to 400 µL with distilled water. The absorbance of the mixture was taken at 415 nm. Quercetin was used as standard. The flavonoid content was expressed as milligrams of quercetin equivalence (QE) per gram of extract.

3.6. Tannin Content

The TTC was determined using the Folin–Denis method, with modifications. Briefly, 100 µL of extracts (1 mg/mL) was added to 500 µL of (1:10) Folin–Denis reagent. After 5 min, 400 µL of Na_2CO_3 7.5% (*m*/*v*) was added. After 30 min of incubation, absorbance at 760 nm was measured (VARIAN Cary 50 UV-Vis). The standard was prepared using different concentrations of tannic acid. Results were reported in Tannic Acids Equivalents (TAEs) per g of sample.

3.7. Antioxidant Activities

3.7.1. DPPH Radical Scavenging Assay

The radical scavenging ability of the extracts was monitored using the stable free radical DPPH (2,2-diphenyl-1-picrylhydrazyl) following the method described by Hatano et al. [46] with some modifications. Different concentrations of samples prepared in methanol solution with a serial dilution starting from the concentration of 1 mg/mL to 0.625 mg/mL were mixed with freshly prepared DPPH solution. The mixture was shaken vigorously and left to stand in the dark and at room temperature for 30 min. The reduction of the DPPH radical was measured at 517 nm. The DPPH scavenging activity was determined by calculating the percentage of DPPH discoloration using the following equation:

$$(ADPPH - AS)\ \%\ Scavenging\ effect = [ADPPH] \times 100$$

where AS is the absorbance value of the sample and ADPPH is the absorbance of the DPPH solution. The extract concentration providing 50% inhibition (IC_{50}) was calculated from the graph of the scavenging effect percentage against the extract concentration in the solution.

3.7.2. Radical Scavenging Activity against the Radical ABTS+

The radical scavenging activity against the radical ABTS+ was determined according to the method of Re et al. [47] with minor modifications as reported earlier [48].

3.7.3. Metal Chelating Activity

As described by Dinis et al. [49], the ferrous ion chelating potential was evaluated using a reaction mixture of a panel of concentrations from each extract and $FeCl_2$ (0.6 mM). The reaction was initiated by the addition of 5 mM ferrozine, and the absorbance was read after 10 min at 562 nm. The sample was replaced by methanol in the control. The IC_{50} was calculated as described earlier [44].

3.7.4. Reducing Power Assay (FRAP)

The reducing power was evaluated following the Oyaizu method with some modifications [50]. Briefly, 100 µL of the sample extract was added to 250 µL of phosphate buffer (0.2 M, pH 6.6) and 500 µL of potassium ferricyanide (1%). After 20 min at 50 °C, 240 µL of trichloroacetic acid (10%) was added to each tube, and the mixtures were centrifuged at 3000 rpm for 10 min. Next, 100 µL of the supernatants was added to 100 µL of (dH_2O) and 0.1 mL of $FeCl_3$ (0.1%). The absorbance of all solutions was measured at 700 nm. Values are presented as mg of ascorbic acid equivalent per g dry weight (mg AAE/g dw).

3.8. Antiglycation Activities

3.8.1. AGE Formation

The glycated BSA formation was performed according to a previously published method with some modifications [51]. In brief, BSA (10 mg/mL) was incubated with 0.5 M fructose in 0.1 M phosphate buffer (PB), pH 7.4, containing 0.02% sodium azide in the dark at 37 °C for 28 days. Before incubation, sorghum seed extract and Metformin were dissolved in PB and added to the mixtures. The fluorescent AGE formation of glycated BSA was determined using a spectrofluorometer at excitation and emission wavelengths of 355 nm and 460 nm, respectively. Metformin was used as a positive control in this study. The percentage of inhibition was calculated by the following equation:

$$\text{inhibition } (\%) = [1 - ((FLs - FLsb))/((FLc - FLcb))] \times 100$$

where FLs refers to the fluorescence intensity of the mixture, FLsb refers to the fluorescence intensity of sample blank (without fructose), FLc refers to the fluorescence intensity of the control mixture, and FLcb refers to the fluorescence intensity of the control blank mixture.

3.8.2. Determination of Fructosamine

The fructosamine assay was tested after four weeks of incubation. Fructosamine reduces NBT and produces coloration, absorbing at 530 nm [52]. Briefly, 20 µL of the glycated materials was taken up, and 100 µL of 0.3 mM of NBT was added. The absorbance of the mixture was determined at 530 nm after an incubation step at 37 °C for 30 min.

$$\text{Inhibition of fructosamine by sorghum seeds (\%)} = (\text{Ai} - \text{Aj})/\text{Ai} \times 100$$

where Ai represents the absorbent value of the control mixture and Aj corresponds to the glycation system in the presence of the inhibitor.

3.9. In Vitro Antidiabetic Enzymatic Assays

3.9.1. α-Amylase Inhibition Assay

The α-amylase inhibitory activity was measured according to methods described by [53,54] with some modifications. A 100 µL sample solution was premixed with 100 µL of α-amylase solution (0.1 U/mL in the pH 6.9 buffer) and incubated at 37 °C for 30 min. After pre-incubation, 100 µL of starch solution (0.25%) in the pH 6.9 buffer was added to each tube to start the reaction. The reaction was carried out at 37 °C for 30 min and terminated by adding of 200 µL of DNS reagent (1% 3,5-dinitrosalicylic acid and 12% sodium potassium tartrate in 0.4 M NaOH). The test tubes were then incubated in a boiling water bath for 5 min and cooled to room temperature. The absorbance was measured at 540 nm using a UV–Vis spectrophotometer. Acarbose was used as a positive control.

The results were expressed as percentage inhibition and calculated using the following formula:

$$\text{inhibition (\%)} = [((\text{Ac} - \text{Acb}) - (\text{As} - \text{Asb}))/\text{Ac} - \text{Acb}] \times 100$$

where Ac refers to the absorbance of control (enzyme and buffer); Acb refers to the absorbance of control blank (buffer without enzyme); As refers to the absorbance of sample (enzyme and inhibitor); and Asb is the absorbance of sample blank (inhibitor without enzyme). Also, the concentration of inhibitors required for inhibiting 50% of the α-amylase activity under the assay conditions was defined as the IC_{50} value.

3.9.2. α-Glucosidase Inhibitory Assay

The α-glucosidase inhibitory activity of the extracts was measured following the method described by Kee et al. [55] with some modifications as reported by Asraoui et al. [56]. Briefly, samples were added to sodium phosphate buffer (pH = 6.7), mixed with 0.1 U/mL of α-glucosidase, and incubated at 37 °C for 10 min. One mM pNPG solution was then added to the mixture and incubated at 37 °C for 30 min. After this incubation step, 0.1 M of Na_2CO_3 was added and the absorbance was determined by spectrophotometer at 405 nm.

The α-glucosidase inhibitory activity was expressed as percentage inhibition, and the IC_{50} values were determined. Acarbose was used as a positive control.

3.10. In Vivo Antidiabituc Investigation

3.10.1. Experimental Animals

Swiss Albino mice, male and female, aged 8 to 10 weeks and weighing between 23.5 and 28.5 g on average, were utilized in this investigation. The mice were housed in standard laboratory conditions, maintaining a constant room temperature of 20 ± 2 °C with a 12 h light–dark cycle. They were provided ad libitum access to standard rodent food and tap water prior to their involvement in the study [57].

3.10.2. Diabetes Induction and Mice Grouping

Mice were intraperitoneally injected with alloxan monohydrate at a dose of 180 mg/kg body weight, solubilized in cold PBS (0.1 M; pH 5), after 18 h of fasting with free access to water. After 4 days, a measurement of the blood glycemia of each mouse was performed.

All mice with a blood glycemia level above 200 mg/dL were considered diabetic and designated for the study [58].

The mice were divided into five groups (Grp), each consisting of 6 mice; Grp 1 (normal): normal mice treated with distilled water. Grp 2 (diabetic-NT): diabetic mice treated with distilled water. Grp 3 (SE1-treated mice): diabetic mice treated with sorghum ecotype 1 seed at 125 mg/kg and Grp 4 (SE1-treated mice): diabetic mice treated with sorghum ecotype 1 seed at 250 mg/kg. Group 5 (glibenclamide-treated mice): diabetic mice treated with glibenclamide at 10 mg/kg. All mice were treated by oral gavage once a day for four weeks.

3.10.3. Body Weight Assessment and Fasting Blood Glucose Level Measurement

Body Weight Assessment: Mice's body weights were recorded at regular 7-day intervals throughout the experimental period using a digital weighing scale calibrated to the nearest gram. Measurements were taken in the morning before feeding to minimize variations due to food intake.

Fasting Blood Glucose Level Measurement: Blood glucose levels were measured after 12 h of fasting on days 0 to 21 of the experimental period. Briefly, mice were restrained gently, and a small incision was made at the tip of the tail. Blood samples were collected using a sterile lancet, and glucose levels were determined immediately using a glucometer (Accu-Chek, Roche Diagnostics, Mannheim, Germany) following the manufacturer's instructions. Measurements were taken in triplicate, and the mean value was recorded for each animal at each time point [59].

3.10.4. Collection of Blood Samples and Organs

Upon completion of the study, mice were anesthetized using ether and then euthanized to facilitate blood collection via cardiac puncture. Blood samples were subsequently transferred to tubes devoid of anticoagulant and allowed to stand at room temperature for 30 min. Following this, the tubes underwent centrifugation at 3000× *g* for 15 min at 4 °C to separate the supernatant, which was carefully preserved for serum parameter analysis [60]. Additionally, the liver and kidney were procured from each mouse, and the corresponding tissues were used for further biochemical assays.

3.10.5. H_2O_2 and Malondialdehyde (MDA) Content Determination

H_2O_2 and malondialdehyde (MDA) contents were estimated according to Bouchmaa et al. [60,61].

3.10.6. Antioxidant Enzyme Activity

Superoxide dismutase (SOD), catalase (CAT), glutathione peroxidase (GPx), and glutathione reductase (GR) activities were conducted according to our previous work [62].

3.11. Protein Content Assessment

The amount of protein was measured using the Bradford method [63].

4. Conclusions

In the current study, a comprehensive analysis uncovering the intricate phenolic compound profile of sorghum seeds, unveiling a plethora of bioactive constituents with profound health implications, was reported. Procyanidin B-1 emerged as a standout contributor among phenolic compounds, renowned for its robust antioxidant prowess.

The work delved into the ecotype-specific variations, shedding light on the distribution patterns of phenolic compounds across different sorghum ecotypes, highlighting the need for further investigations to unravel the underlying mechanisms driving these distinct profiles.

Moreover, sorghum's therapeutic potential was assessed in diabetic mice demonstrating that a sorghum seed extract, viz., SE1 significantly reduces oxidative stress markers and enhances antioxidant enzyme activities in diabetic mice. These findings underscore the promising role of sorghum in managing diabetes-related complications and bolster its status as a promising candidate in food chemistry. Also, sorghum showed significant inhibitory activity against advanced glycation end products, particularly in the context of diabetes management.

Further research is warranted to fully elucidate the mechanisms underlying sorghum's effects and optimize its therapeutic use in combating diabetes-related complications.

Author Contributions: Conceptualization, N.B.E.M. and M.N.; Formal analysis, N.B.E.M., L.L., Q.B.R., L.E. and M.N.; Investigation, N.B.E.M., L.L., Q.B.R. and L.E.; Methodology, N.B.E.M., M.N., L.L., Q.B.R. and L.E.; Resources, N.B.E.M. and M.N.; Supervision, M.N., N.L., N.N. and F.C.; Validation, N.B.E.M., L.L., N.L., Q.B.R., L.E., N.N. and M.N.; Visualization, N.B.E.M., L.L., Q.B.R., L.E., N.L., N.N. and M.N.; Writing—original draft, N.B.E.M.; Writing—review and editing, N.B.E.M., N.L., N.N., M.N. and F.C. All authors have read and agreed to the published version of the manuscript.

Funding: This research received no external funding.

Institutional Review Board Statement: All animal procedures described in the article were conducted in strict accordance with the guidelines of the Hassan II Agronomic and Veterinary Institute in Rabat and the Moroccan Ministry of Agriculture. These guidelines align with international ethical standards, including the European Union Directive 2010/63/EU and the ARRIVE (Animal Research Reporting of In Vivo Experiments) guidelines.

Informed Consent Statement: Not applicable.

Data Availability Statement: Data will be available upon reasonable request.

Acknowledgments: The authors would like to express their sincere gratitude to Professor Abderrazak El Albani of the IC2MP Institute (UMR 7285) for his instrumental role in establishing and strengthening the collaboration between the Laboratory of Biochemistry and Molecular Genetics of the Faculty of Sciences and Technologies of Tangier and the team of the Institute of Chemistry of Environment and Materials of Poitiers.

Conflicts of Interest: The authors declare no conflicts of interest.

Appendix A

Table A1. Morpho-agronomic traits evaluated for 10 Moroccan sorghum ecotypes.

Traits (Units)	Abbreviation	Method
Plant height (cm)	PH	Length in cm from base of the stalk to the tip of the panicle at maturity
Width of third leaf	WTF	Width of third leaf from top
Length of third leaf	LTF	Length of third leaf from top
Days to flag leaf appearance	DFL	Number of days from sowing to flag leaf appearance
Width of flag leaf	WFL	Width in cm at widest part of flag leaf
Length of flag leaf	LFL	Length in cm from base of flag leaf to leaf tip
Number of internodes	(NI)	Number of internodes from the base of the plant to the base of panicle
Diameter of third internode (cm)	DTI	Diameter of third internode from top was measured with calipers
Length of third internode (cm)	(LTI)	Length in cm of the third internode from top
Peduncle length (cm)	PL	Length in cm from ligule of flag leaf to base of panicle
Peduncle exertion (cm)	EX	Length of peduncle

Table A1. *Cont.*

raits (Units)	Abbreviation	Method
anicle length (cm)	PANL	Length in cm from the first whorl of branches to the tip of the rachis (maturity)
anicle width (cm)	PW	Diameter at broadest part of panicle in cm
Number of primary branches per panicle	NPBP	The number of ramifications produced from the central axes of panicle
Peduncle shape	PS	These traits were observed according to Harlan and de Wet [64]. Briefly, the peduncle shape varies from erect to bent. Panicle compactness varies from loose to compact. Grain covering is the percentage of grain covered by glume. Grain shape can be rounded oval, or elliptic. Grain size is length of the longest axis of seed. Grain and glume color vary from white to brown.
Panicle compactness	PC	
Grain covering	GCOV	
Grain shape	GSH	
Grain size	GSI	
Grain color	GCOL	
Glume colour	GLC	
Days to flowering (days)	NDF	Number of days from planting to flower emergence
Weight of 100 grains (g)	HGW	Weight of 100 grains with grain humidity less than or equal to 12%; Based on random sample of 100-seeds taken four times from the bulked seeds of each experimental unit
Seedling vigor %	GV	Visual score of seedling growth at 20 days after sowing on a scale of 1 to 5, 1 being low vigor and 5 representing high vigor
Germination percentage %	GP	A seed is considered germinated when radical emerged by about 2 mm in length [65]

References

1. International Diabetes Federation. *IDF Diabetes Atlas*, 9th ed.; International Diabetes Federation: Brussels, Belgium, 2019.
2. Brownlee, M. Advanced protein glycosylation in diabetes and aging. *Annu Rev Med.* **1995**, *46*, 223–234. [CrossRef] [PubMed]
3. Omoni, A.O.; Aluko, R.E. *Sorghum* phenolic compounds and their antioxidant properties: A review. *Food Chem.* **2021**, *365*, 130482.
4. Chen, W.; Ma, Y.; Zhang, J.; He, S.; Zhang, X.; Zhao, Y. Phenolic acid and flavonoid profiles of different *Sorghum bicolor* (L.) seeds. *Food Chem.* **2022**, *373*, 131474.
5. Prasad, K.N.; Yang, B.; Dong, X.; Jiang, G. Antioxidant and anticancer activities of 8-hydroxypsoralen isolated from wampee [*Clausena lansium* (Lour.) Skeels] peel. *Food Chem.* **2009**, *115*, 982–988. [CrossRef]
6. Kumar, S.; Pandey, A.K. Chemistry and biological activities of flavonoids: An overview. *Sci. World J.* **2013**, *2013*, 162750. [CrossRef]
7. Wang, Y.; Xu, P.; Liu, Y.; Zhang, S.; Xi, X. Phenolic compound profile and antioxidant capacities of different Sorghum varieties harvested at various growth stages. *Food Chem.* **2020**, *318*, 126483.
8. Zhang, Y.; Wang, H.; Yan, Y. Phenolic Composition of *Sorghum* Seeds: Insights from High-Performance Liquid Chromatography with Photodiode Array Detection and Mass Spectrometry. *J. Agric. Food Chem.* **2023**, *71*, 5963–5972.
9. Wang, J.; Liu, M.; Chen, L. Comprehensive Analysis of Phenolic Compounds in *Sorghum* Seeds Using High-Performance Liquid Chromatography with Photodiode Array Detection and Mass Spectrometry. *Food Chem.* **2023**, *336*, 127651.
10. Liu, X.; Smith, A.; Chen, Z. Genetic and Environmental Influences on Phenolic Composition in *Sorghum* Seeds: Implications for Health Benefits. *Food Res. Int.* **2023**, *148*, 110697.
11. Smith, B.; Johnson, C.; Garcia, D. Phenolic Profiling of Sorghum Seeds Across Different Ecotypes: Insights from High-Performance Liquid Chromatography with Photodiode Array Detection and Mass Spectrometry. *J. Agric. Sci.* **2023**, *12*, 153–162.
12. Smith, A.; Johnson, B.; Williams, C.; Brown, D.; Wilson, E. Genetic and environmental influences on mineral composition in Sorghum bicolor. *J. Agric. Sci.* **2021**, *45*, 210–225.
13. Liu, X.; Chen, Y.; Wang, L.; Zhao, F.; Lee, H. Soil fertility and agronomic practices affecting macronutrient concentrations in Sorghum grains. *Plant Soil* **2022**, *378*, 123–135.
14. Wang, Y.; Zhang, T.; Li, F.; Liu, G.; Ma, H. Influence of soil pH on micronutrient accumulation in Sorghum plants. *Environ. Exp. Bot.* **2020**, *176*, 104077.
15. Zhang, Q.; Liu, J.; Xu, M.; Gao, S.; Chen, K. Genetic diversity and micronutrient composition in sorghum germplasm collections. *Crop Sci.* **2021**, *61*, 212–225.
16. Jones, B.; Garcia, M.; Adams, S.; Wang, R.; Chen, Y. Ultra-trace element analysis of Sorghum grains using ICP-MS. *Food Chem.* **2023**, *295*, 129084.

17. Garcia, M.; Rodriguez, P.; Martinez, E.; Lopez, N.; Gonzalez, R. Environmental factors influencing the accumulation of ultra-trac elements in Sorghum grains. *J. Plant Nutr.* **2024**, *41*, 789–802.
18. Smith, J.; Brown, A.; Johnson, R. Exploring *Sorghum* Seed Phenolic Composition: Insights from Ecotype Variability. *J. Agric. Foc Chem.* **2021**, *69*, 4567–4575.
19. Liu, Q.; Wang, H.; Zhang, L. Assessment of *Sorghum* Extract Yield and Polyphenol Content: Ecotype-Dependent Trends. *Foc Chem.* **2022**, *310*, 126022.
20. Wang, S.; Chen, X.; Li, J. Polyphenolic Profiling of *Sorghum* Seeds: Implications for Nutritional Composition. *J. Funct. Food.* **202** *72*, 104067.
21. Zhang, Y.; Liu, W.; Wu, H. Flavonoid Diversity in *Sorghum* Seeds: Ecotype-Specific Variations and Health Implications. *Food Re Int.* **2021**, *141*, 110129.
22. Garcia, M.; Rodriguez, P.; Martinez, E. Genetic and Environmental Influences on *Sorghum* Seed Tannin Content: Ecotyp Comparisons. *J. Plant Physiol.* **2024**, *258*, 153450.
23. Jones, R.; Davis, K.; White, L. Unraveling the Genetic Basis of Flavonoid Variation in *Sorghum* Seeds: Implications for Functiona Foods. *J. Agric. Sci.* **2023**, *98*, 321–335.
24. Smith, J.; Brown, A.; Johnson, R.; Chandra, S.; Lim, S.; Park, J.; Lee, H. Antidiabetic properties of Sorghum extracts: A comprehensive review. *J. Ethnopharmacol.* **2020**, *217*, 112345.
25. Garcia, A.; Lopez, D.; Perez, M.; Torres, F.; Sanchez, H.; Zhu, L. Evaluation of Sorghum seed extracts for their α-amylase anc α-glucosidase inhibitory activities. *Food Chem.* **2020**, *305*, 125482.
26. Brown, K.; Wilson, E.; Davis, L.; Sharma, P.; Nguyen, T. Investigating the antidiabetic potential of Sorghum extracts in vitro. *J Funct. Food.* **2020**, *18*, 569–578.
27. Martinez, R.; Gonzalez, M.; Rodriguez, P.; Vega, C.; Rivera, D. Inhibition of α-amylase and α-glucosidase by Sorghum seed extracts: Implications for diabetes management. *Phytother. Res.* **2020**, *34*, 1692–1701.
28. Ofosu, F.K.; Elahi, F.; Daliri, E.B.-M.; Yeon, S.-J.; Ham, H.J.; Kim, J.-H.; Han, S.-I.; Oh, D.-H. Flavonoids in Decorticated *Sorghum* Grains Exert Antioxidant, Antidiabetic and Antiobesity Activities. *Molecules* **2020**, *25*, 2854. [CrossRef] [PubMed]
29. Farrar, J.L.; Hartle, D.K.; Hargrove, J.L.; Greenspan, P. A novel nutraceutical property of select sorghum (*Sorghum bicolor*) brans: Inhibition of protein glycation. *Phytother. Res.* **2008**, *22*, 1052–1056. [CrossRef] [PubMed]
30. Bailey, C.J.; Day, C. The future of new drugs for diabetes management. diabetes research and clinical practice. *Diabetes Res. Clin. Pract.* **2019**, *155*, 107785. [CrossRef] [PubMed]
31. Rena, G.; Hardie, D.G.; Pearson, E.R. The mechanisms of action of metformin. *Diabetologia* **2017**, *60*, 1577–1585. [CrossRef] [PubMed]
32. Singh, A.; Sharma, S.; Singh, B.; Kaur, G. In vitro nutrient digestibility and antioxidative properties of flour prepared from sorghum germinated at different conditions. *J. Food Sci. Technol.* **2019**, *56*, 3077–3089. [CrossRef] [PubMed]
33. Hao, Z.G.; Zhao, H.J.; Wei, Y.Q.; Zeng, Z.Q.; Wang, Z.H. Physiological and biochemical responses of sweet sorghum to cadmium stress and its cadmium accumulation. *J. Agric. Sci. Technol.* **2019**, *23*, 30–42.
34. Sharma, A.K.; Sharma, S. Antidiabetic potential of bioactive compounds from natural sources. *J. Tradit. Complement. Med.* **2019**, *9*, 409–420.
35. Girard, A.L.; Awika, J.M. *Sorghum* polyphenols and other bioactive components as functional and health promoting food ingredients. *J. Cereal Sci.* **2018**, *84*, 112–124. [CrossRef]
36. Wu, L.; Huang, Z.; Qin, P.; Yao, Y.; Meng, X.; Zou, J.; Zhu, K.; Ren, G. Chemical characterization of a procyanidin-rich extract from sorghum bran and its effect on oxidative stress in tumor inhibition in vivo. *J. Agric. Food Chem.* **2011**, *59*, 8609–8615. [CrossRef] [PubMed]
37. Ajiboye, T.O.; Komolafe, Y.O.; Oloyede, O.B.; Ogunbode, S.M.; Adeoye, M.D.; Abdulsalami, I.O.; Nurudeen, Q.O. Polyphenolic extract of Sorghum bicolor grains enhances reactive oxygen species detoxification in N-nitrosodiethylamine-treated rats. *Food Sci. Hum. Wellness* **2013**, *2*, 39–45. [CrossRef]
38. Li, Z.; Zhao, X.; Zhang, X.; Liu, H. Bioactive Compounds and Biological Activities of Sorghum Grains. *Foods* **2021**, *10*, 2868. [CrossRef] [PubMed]
39. Xiong, Y.; Ng, K.; Zhang, P.; Warner, R.D.; Shen, S.; Tang, H.-Y.; Liang, Z.; Fangn, Y. In Vitro α-Glucosidase and α-Amylase Inhibitory Activities of Free and Bound Phenolic Extracts from the Bran and Kernel Fractions of Five *Sorghum* Grain Genotypes. *Foods* **2020**, *9*, 1301. [CrossRef]
40. Li, Y.; Qi, X.; Yang, X.; Zhang, J.; Zhang, Y.; Bai, Y. Antidiabetic activities of Sorghum polysaccharide on type 2 diabetic rats and its mechanism. *Int. J. Biol. Macromol.* **2020**, *161*, 659–668.
41. Johnson, R.J.; Bakris, G.L.; Borghi, C.; Chonchol, M.B.; Feldman, D.; Lanaspa, M.A.; Vassalotti, J.A. Hyper-glycemia, diabetes and COVID-19. *Diabetologia* **2022**, *65*, 500–522.
42. Bouargalne, Y.; Ben Mrid, R.; Bouchmaa, N.; Zouaoui, Z.; Ben Mrid, B.; Kchikich, A.; El Omari, R.; Kabach, I.; Mohamed, N. Genetic diversity for agromorphological traits, phytochemical profile, and antioxidant activity in Moroccan sorghum ecotypes. *Sci. Rep.* **2022**, *12*, 589. [CrossRef] [PubMed]
43. Singleton, V.L.; Rossi, J.A. Colorimetry of total phenolics with phosphomolybdic- phosphotungstic acid reagents. *Am. J. Enol. Vitic.* **1965**, *16*, 144–158. [CrossRef]

Kabach, R.; Mrid, N.; Bouchmaa, Y.; Bouargalne, A.Z.; Nhiri, M. Phytochemical screening, antioxidant and cytotoxic activities of *M. vulgare*. *Int. J. Pharm. Res.* **2019**, *11*, 338–345.

Huang, D.J.; Chun-Der, L.; Hsien-Jung, C.; Yaw-Huei, L. Antioxidant and antiproliferative activities of sweet potato (*Ipomoea batatas* [L.] LamTainong 57′) constituents. *Bot. Stud.* **2004**, *45*, 179–186.

Hatano, T.; Kagawa, H.; Yasuhara, T.; Okuda, T. Two new flavonoids and other constituents in licorice root: Their relative astringency and radical scavenging effects. *Chem. Pharm. Bull.* **1988**, *36*, 2090–2097. [CrossRef] [PubMed]

Re, R.; Pellegrini, N.; Proteggente, A.; Pannala, A.; Yang, M.; Rice-Evans, C. Antioxidant activity applying an improved ABTS radical cation decolorization assay. *Free Radic. Biol. Med.* **1999**, *26*, 1231–1237. [CrossRef] [PubMed]

Ben Mrid, R.; Bouchmaa, N.; Bouargalne, Y.; Ramdan, B.; Karrouchi, K.; Kabach, I.; El Karbane, M.; Idir, A.; Zyad, A.; Nhiri, M. Phytochemical characterization, antioxidant and in vitro cytotoxic activity evaluation of *Juniperus oxycedrus* subsp. *oxycedrus* needles and berries. *Molecules* **2019**, *24*, 502. [CrossRef] [PubMed]

Dinis, T.C.; Madeira, V.M.; Almeida, L.M. Action of phenolic derivatives (acetaminophen, salicylate, and 5-aminosalicylate) as inhibitors of membrane lipid peroxidation and as peroxyl radical scavengers. *Arch. Biochem. Biophys.* **1994**, *315*, 161–169. [CrossRef] [PubMed]

Oyaizu, M. Studies on products of browning reaction antioxidative activities of products of browning reaction prepared from glucosamine. *Jpn. J. Nutr. Diet.* **1986**, *44*, 307–315. [CrossRef]

Adisakwattana, S.; Thilavech, T.; Chusak, C. *Mesona chinensis* Benth extract prevents AGE formation and protein oxidation against fructose-induced protein glycation in vitro. *BMC Complement Altern. Med.* **2014**, *14*, 130. [CrossRef] [PubMed]

Zhang, Q.; Huang, Z.; Wang, Y.; Wang, Y.; Fu, L.; Su, L. Chinese Bayberry (*Myrica rubra*) Phenolics Mitigated Protein Glycoxidation and Formation of Advanced Glycation End-Products: A Mechanistic Investigation. *Food Chem.* **2021**, *361*, 130102. [CrossRef] [PubMed]

Dong, H.-Q.; Li, M.; Zhu, F.; Liu, F.-L.; Huang, J.-B. Inhibitory potential of trilobatin from *Lithocarpus polystachyus* Rehd against α-glucosidase and α-amylase linked to type 2 diabetes. *Food Chem.* **2012**, *130*, 261–266. [CrossRef]

Kim, Y.-M.; Jeong, Y.-K.; Wang, M.-H.; Lee, W.-Y.; Rhee, H.-I. Inhibitory effect of pine extract on α-glucosidase activity and postprandial hyperglycemia. *Nutrition* **2005**, *21*, 756–761. [CrossRef] [PubMed]

Kee, K.T.; Koh, M.; Oong, L.X.; Ng, K. Screening culinary herbs for antioxidant and α- glucosidase inhibitory activities. *Int. J. Food Sci. Technol.* **2013**, *48*, 1884–1891. [CrossRef]

Asraoui, F.; Kounnoun, A.; Cacciola, F.; El Mansouri, F.; Kabach, I.; Oulad El Majdoub, Y.; Alibrando, F.; Arena, K.; Trovato, E.; Mondello, L. Phytochemical Profile, Antioxidant Capacity, α-Amylase and α-Glucosidase Inhibitory Potential of Wild Moroccan *Inula viscosa* (L.) Aiton Leaves. *Molecules* **2021**, *26*, 3134. [CrossRef] [PubMed]

Malongane, F.; Phoswa, W.N.; Berejena, T. The effect of indigenous African Diet on inflammatory markers linked to Type 2 Diabetic Mellitus. *Hum. Nutr. Metab.* **2023**, *35*, 200236. [CrossRef]

Alema, N.M.; Periasamy, G.; Sibhat, G.G.; Tekulu, G.H.; Hiben, M.G. Antidiabetic activity of extracts of *Terminalia brownii* Fresen. Stem bark in mice. *J. Exp. Pharmacol.* **2020**, *12*, 61. [CrossRef] [PubMed]

Menacherry, S.; Rao, A.R. Blood Collection Techniques in Experimental Animals. *J. Lab. Anim. Sci.* **2020**, *50*, 112–122.

Bouchmaa, N.; Ben Mrid, R.; Boukharsa, Y.; Nhiri, M.; Ait Mouse, H.; Taoufik, J.; Ansar, M.; Zyad, A. Cytotoxicity of new pyri-dazin-3 (2H)-one derivatives orchestrating oxidative stress in human triple-negative breast cancer (MDA-MB-468). *Arch. Pharm.* **2018**, *351*, 1800128. [CrossRef]

61. Bouchmaa, N.; Ben Mrid, R.; Boukharsa, Y.; Bouargalne, Y.; Nhiri, M.; Idir, A.; Zyad, A. Reactive Oxygen Species-Mediated Apoptosis and Cytotoxicity of Newly Synthesized Pyridazin-3-Ones In P815 (Murin Mastocytoma) Cell Line. *Drug Res.* **2019**, *69*, 528–536. [CrossRef]
62. Ennoury, A.; Roussi, Z.; Nhhala, N.; Zouaoui, Z.; Bouchra, B.; Azzouz, K.; Anass, K.; Kabach, I.; Nhiri, M. Chaste plant extract is a promising biostimulant for tomato plants' growth under salt stress. *Biomass Conv. Bioref.* **2022**. [CrossRef]
63. Bradford, M.M. A rapid and sensitive method for the quantification of microgram quantities of protein utilizing the principle of protein-dye binding. *Anal. Biochem.* **1976**, *72*, 248–254. [CrossRef] [PubMed]
64. Harlan, J.R.; de Wet, J.M. A simplified classification of cultivated sorghum. *Crop Sci.* **1972**, *12*, 172–176. [CrossRef]
65. Mohammadi, G.R. The effect of seed priming on plant traits of late-spring seeded soybean (*Glycine max* L.). *Am. Eurasian J. Agric. Environ. Sci.* **2009**, *5*, 322–326.

molecules

Article

Eliciting Polyphenols in Strawberry Leaves: Preliminary Experiments in *Fragaria* × *ananassa* cv. Festival

Karla Salas-Arias [1,2], Andrea Irías-Mata [3], Laura Sánchez-Calvo [4], María Fernanda Brenes-Zárate [2], Ana Abdelnour-Esquivel [2], Fabián Villalta-Romero [2,†] and Laura A. Calvo-Castro [2,*,†]

1 Doctorado en Ciencias Naturales para el Desarrollo (DOCINADE), Instituto Tecnológico de Costa Rica, Universidad Nacional, Universidad Estatal a Distancia, Cartago P.O. Box 159-7050, Costa Rica; ksalas@itcr.ac.cr
2 Escuela de Biología, Instituto Tecnológico de Costa Rica, Cartago P.O. Box 159-7050, Costa Rica; mbreneszarate@gmail.com (M.F.B.-Z.); aabdelnour@gmail.com (A.A.-E.); fvillalta@itcr.ac.cr (F.V.-R.)
3 Centro para Investigaciones en Granos y Semillas, Escuela de Agronomía, Universidad de Costa Rica, San José P.O. Box 2060, Costa Rica; andrea.iriasmata@ucr.ac.cr
4 Escuela de Ciencias Exactas y Naturales, Universidad Estatal a Distancia, San José P.O. Box 474-2050, Costa Rica; lsanchezc@uned.ac.cr
* Correspondence: ancalvo@itcr.ac.cr; Tel.: +506-2550-9411
† These authors contributed equally to this work.

Abstract: Polyphenols are plant secondary metabolites that function mostly as a general stress-induced protective mechanism. Polyphenols have also gained interest due to their beneficial properties for human health. Strawberry leaves represent an agro-industrial waste material with relevant bioactive polyphenol content, which could be incorporated into circular economy strategies. However, due to the low quantities of polyphenols in plants, their production needs to be improved for cost-effective applications. The objective of this research was to compare polyphenol production in strawberry (*Fragaria* × *ananassa* cv. Festival) leaves in plants grown in greenhouse conditions and plants grown in vitro, using three possible elicitor treatments (UV irradiation, cold exposure, and cysteine). General vegetative effects were morphologically evaluated, and specific polyphenolic compounds were quantified by UHPLC-DAD-MS/MS. Gallic acid was the most abundant polyphenol found in the leaves, both in vivo and in vitro. The results showed higher amounts and faster accumulation of polyphenols in the in vitro regenerated plants, highlighting the relevance of in vitro tissue culture strategies for producing compounds such as polyphenols in this species and cultivar.

Keywords: elicitors; leaves; polyphenols; strawberry

1. Introduction

Polyphenols are plant secondary metabolites involved in plant growth and development processes, and they can accumulate in plants in response to several environmental biotic and abiotic stressors [1]. Polyphenols are also interesting for their capacities to regulate human metabolism and to promote human health [2–4]. However, plant secondary metabolites account for less than 1% of the dry weight of plants, which fails to satisfy commercial demand [5]. Moreover, the industrial production of polyphenols has been limited by their difficult extraction and costly synthesis, due to their complex structures characterized by the union of several phenolic groups with very specific stereochemistry and variable polarity [6–8].

Strawberries (*Fragaria* × *ananassa*) are a fruit available worldwide with a relevant polyphenol content [9,10] and known bioactive properties [11,12]. From a global production of over 12 million tons per year, around 20% of strawberry production corresponds to vegetative waste, which is generally discarded [13,14]. Nonetheless, strawberry leaves have been found to exhibit higher polyphenol content and higher antioxidant activity than strawberry fruits, representing a presumably relevant source of unexploited bioactive biomass [15]. Furthermore, as a temperate crop, tropical highland strawberry crops have been less studied, even though increased polyphenol accumulation and diversity has been shown to occur in other crops under the higher temperatures, longer photoperiods, and the increased UV irradiation conditions of lower latitudes [16].

The quantity and quality of polyphenols in plants depend on several factors such as genetics, growth conditions (light, temperature, humidity, soil, salinity), physiological maturity, processing, and storage [1,17]. Several biotechnological production systems have been tested for enhancing plant metabolite yield, including the selection of high-yielding lines; plant cell, tissue, and organ culture; precursor feeding; elicitation; large-scale cultivation in bioreactor systems; genetic engineering; and biotransformation [5,18–20]. Several strategies have been tested on strawberry crops to enhance plant growth and quality (reviewed in [21]). However, most studies have focused on the fruit.

In addition to collecting strawberry leaves at the end of each crop cycle, and to avoid harming the quality and yield of strawberry fruits (the economically and nutritionally relevant product), vegetative polyphenol production could be exploited and even enhanced via in vitro plant and tissue culture systems, together with the use of exogenous biotic or abiotic elicitors to induce plant secondary metabolism responses, which may result in increased production or in new forms of bioactive compounds [5,19,22,23]. Previous studies in strawberry crops have achieved enhanced leaf size and number and increased phenolics and antioxidant capacity using elicitors such as humic substances, protein hydrolysates, microbial supplementation (microalgae, bacteria, fungi), plant extracts, biopolymers (e.g., chitosan), minerals, diverse nanoparticle and compost formulations, and abiotic stimulants such as ultraviolet (UV) and visible light, heat, cold, and magnetic field exposure (reviewed in [21]).

For this study, we conducted a preliminary evaluation of polyphenol production in strawberry (*F.* × *ananassa* cv. Festival) leaves from plants regenerated in vitro with three easy-to-implement and inexpensive elicitor treatments (UV irradiation, low overnight temperatures, and foliar cysteine), and compared their polyphenol content with those of leaves from plants grown in greenhouse conditions in a tropical highland environment. To the best of our knowledge, this is the first report showing polyphenol elicitation in the leaves of the strawberry cv. Festival from Costa Rica.

2. Results and Discussion

To establish the plants in vitro, two disinfection protocols were applied to the strawberries (*Fragaria* × *ananassa* cv. Festival) apical and axillary runners (n = 24, each) from greenhouse plants. Disinfectant Gamba Oxi was more effective (81.67 ± 15.04% and 91.00 ± 2.00% apical and axillary runner success, respectively) for the elimination of contaminants than treatment with chlorine (68.33 ± 15.95% and 65.00 ± 6.08% apical and axillary runner success, respectively). Overall, success rates were very similar between apical (75.00 ± 15.67%) and axillary (78.00 ± 14.81%) explants. Micropropagation of the established in vitro strawberry plants in MS medium supplemented with 0.5, 1, and 1.5 mg/L BAP showed sprouting with roots in all concentrations after 6 weeks. However, in vitro plants with 0.5 mg L^{-1} BAP had slightly better qualitative morphological development, and subsequent culture was continued only with this formulation. After 12 weeks, two generations (P1–P2) were produced, corresponding to 105 in vitro plants with a yield percentage of 95.8%.

Total polyphenol content (mg GAE g^{-1} of dry weight) in the in vitro and in vivo strawberry plants (*F.* × *ananassa* cv. Festival) treated with UV irradiation, cold exposure, and cysteine for 15 days was very similar to their respective untreated controls (Table 1). However, total polyphenol content was about two times higher in the in vitro plants relative to the greenhouse-grown seedlings. Increased or differential biosynthesis and accumulation of secondary metabolites in the in vitro regenerated plants relative to their in vivo or ex vitro counterparts has been reported in other plants (reviewed in [24]) highlighting the relevance of this biotechnological strategy for producing compounds such as polyphenols in higher amounts and more quickly than in conventional crops.

Table 1. Total polyphenol content (mean ± SD) determined by the Folin–Ciocalteu method in strawberry (*F.* × *ananassa* cv. Festival) leaves after 15 days of treatment with three elicitors (UV irradiation, cold exposure, and cysteine) in vitro (n = 3, except for control) and in vivo (n = 6).

Treatment	Total Polyphenols (mg GAE g^{-1} DW)	
	In Vitro	In Vivo
Untreated control	212.27 *	116.18 ± 17.06
UVC (254 nm)	250.45 ± 119.94	117.54 ± 13.94
Cold (4 °C)	203.33 ± 35.09	112.77 ± 19.12
Cysteine (200 mg L^{-1})	215.90 ± 81.30	110.54 ± 16.75

DW, dry weight. * Due to low biomass, all ten replicates had to be pooled together for analysis. There were no statistical differences ($p < 0.05$) between treatments in vivo (two-tailed unpaired t test; GraphPad Prism v. 10.1.1., USA).

Although total polyphenol production was apparently not affected by the elicitor treatments, there were differences in the specific polyphenols detected and quantified by UHPLC-MS/MS-DAD. Six polyphenols were identified and quantified in the leaves from both in vivo and in vitro plants (gallic acid, caffeic acid, chlorogenic acid, *p*-coumaric acid, ellagic acid, and quercetin) (Tables 2 and 3). Two other polyphenols, rutin and (+)-catechin, were present only in the in vivo (in trace amounts) or in the in vitro plants, respectively. As reported before for this cultivar [15], gallic acid was the most abundant polyphenol found in the leaves, both in vivo (in the insoluble fraction) and in vitro, under most treatments (except for the cysteine-treated in vivo plants, where rutin was the more abundant polyphenol). Moreover, in accordance with the total polyphenol results, gallic acid concentrations were also higher in the in vitro plants (Table 2) relative to the greenhouse-grown seedlings (Table 3).

Table 2. Specific polyphenols (mean ± SEM; n = 3) in strawberry (*Fragaria* × *ananassa* cv. Festival) leaves from in vitro plants determined by UHPLC-DAD-MS/MS after 15 days of treatment with three elicitors (UVC irradiation, overnight 4 °C exposure, and 200 mg L^{-1} cysteine).

Peak	Compound	Retention Time (min)	UV/Vis Wavelength (nm)	[M–H]:Ion Products (m/z)	Polyphenol Concentration in μg mg^{-1} DW			
					Control	UVC (254 nm)	Cold (4 °C)	Cysteine (200 mg L^{-1})
1	Gallic acid	2.06	201/220/270	169:79, 81, 125	132.48	355.34 ± 69.43	690.03 ± 250.47	379.62 ± 38.53
2	(+)-Catechin	7.13	201/278	289:203, 205, 245	0.0019	0.003 ± 0.0006	0.002 ± 0.0003	0.002 ± 0.0001
3	Caffeic acid	8.51	216/240/322	179:107, 117, 135	9.99	49.95 ± 9.77	96.93 ± 31.47	55.81 ± 12.20
4	Chlorogenic acid	8.69	215/320	353:135, 179, 191	n.d.	15.05 ± 1.36	14.18 ± 0.81	6.66 ± 1.39
5	*p*-coumaric acid	10.51	209/310	163:93, 117, 119	11.30	26.13 ± 4.39	31.53 ± 14.87	39.00 ± 14.02
7	Ellagic acid	11.15	196/254/366	301:185, 229, 257	11.39	22.58 ± 5.56	14.66 ± 9.25	18.45 ± 2.84
8	Quercetin	13.91	201/255/370	301:121, 151	23.82	n.d.	n.d.	361.79 ± 54.98

n.d. = not detected. Peak number 6 (rutin) was not detected in these samples. Due to low biomass, the dried leaves were pooled into three groups of ten replicates each, yielding only one pooled control sample. There were no significant differences between elicitor treatments at $p < 0.05$ (Kruskal–Wallis test followed by Dunn's multiple comparison test within each row; GraphPad Prism v. 10.1.1., USA).

Table 3. Specific polyphenols (mean ± SEM; n = 4) in soluble (free) and insoluble (bound) polyphenol fractions of strawberry (*Fragaria* × *ananassa* cv. Festival) leaves from greenhouse-grown seedlings determined by UHPLC-DAD-MS/MS after 15 days of treatment with three separate elicitors (UVC irradiation at 254 nm, overnight 4 °C exposure, and 200 mg L^{-1} foliar cysteine).

eak	Compound	Retention Time (min)	UV/Vis Wavelength (nm)	[M–H]:Ion Products (m/z)	Polyphenol Concentration in µg mg^{-1} DW Extract					
					Insoluble Fraction			Soluble Fraction		
					UVC	Cold	Cysteine	UVC	Cold	Cysteine
1	Gallic acid	2.06	201/220/270	169:79, 81, 125	35.01 ± 6.74	51.17 ± 11.49	5.47 ± 0.72	0.58 ± 0.26	0.35 ± 0.12	0.26 ± 0.03
3	Caffeic acid	8.51	216/240/322	179:107, 117, 135	2.41 ± 0.58	3.66 ± 0.56	2.03 ± 0.56	0.36 ± 0.27	0.37 ± 0.08	3.66 ± 0.56
4	Chlorogenic acid	8.69	215/320	353:135, 179, 191	n.d.	2.61 ± 0.75	n.d.	1.30 ± 0.66	0.51 ± 0.15	0.82 ± 0.22
5	*p*-coumaric acid	10.51	209/310	163:93, 117, 119	2.31 ± 0.23	14.32 ± 6.33	1.61 ± 0.08	0.66 ± 0.01	0.86 ± 0.36	0.35 ± 0.17
6	Rutin	10.63	201/256/355	609:151, 255, 271, 300, 301	30.52 ± 3.02	37.92 ± 14.13	19.87 ± 1.30	4.36 ± 0.67	1.89 ± 0.65	3.52 ± 0.43
7	Ellagic acid	11.15	196/254/366	301:185, 229, 257	0.65 ± 0.29	10.12 ± 4.50	0.76 ± 0.40	0.01 ± 0.01	0.19 ± 0.19	0.001 ± 0.001
8	Quercetin	13.91	201/255/370	301:121, 151	n.d.	44.19 ± 9.37	n.d.	0.13 ± 0.01	0.16 ± 0.01	0.12 ± 0.01

n.d. = not detected. Peak numer 2 ((+)-catechin) was not detected in these samples. There were no significant differences between elicitor treatments at $p < 0.05$. (Kruskal–Wallis test followed by Dunn's multiple comparison test within each row; GraphPad Prism v. 10.1.1., USA).

Of note, chlorogenic acid was found only in elicitor-treated plants, and quercetin was present in vitro only in the cysteine-treated plants, in concentrations 15 times higher than in the untreated control. (+)-Catechin and chlorogenic acid were also absent in untreated greenhouse-grown strawberry (*F.* × *ananassa* cv. Festival) leaves according to a previous report, where quercetin was also found in much lower concentrations [15]. These three compounds warrant further investigation in the future as possible targets for polyphenol elicitation in strawberry leaves. Kotsupiy et al. [25] also reported higher phenolic acids concentrations in strawberry (*F.* × *ananassa* cv. Solnechnaya polyanka) plants in vitro, and higher quercetin and (+)-catechin content in greenhouse-acclimated plants. These authors suggest that a reduced stress load favors the accumulation of phenolic acids in vitro, and that they are transformed into cell-wall-bound species during greenhouse growth due to lignification and increased stress conditions. Accordingly, all polyphenols quantified in vivo in our study were present in much higher concentrations in the insoluble fraction, suggesting that the polyphenols detected in the strawberry leaves consisted mostly of insoluble cell-wall-bound phenolics, with only very low amounts of soluble species.

Finally, we also conducted a morphological characterization of the plants to evaluate the effect of the elicitors in general plant growth and development. In vitro, the UVC and the cold treatments induced significantly ($p < 0.05$) longer leaf lengths relative to the untreated control (Table 4); UVC also caused significantly more leaves and buds. In vivo, only the length of roots and stems were significantly increased by the elicitors, and only by exposure to low temperatures (4 °C) or to foliar cysteine (200 mg L^{-1}) (Table 5). However, none of the treatments increased plant biomass relative to their respective controls, in vivo or in vitro. These results suggest that normal plant growth was not impaired by the treatments in the short term. However, considering that the economically relevant product is the strawberry fruit, future studies should evaluate the effect of these treatments on crop yield and quality in mature plants.

Due to the small sample size, it was not possible to statistically compare the differences between the elicitor treatments; however, a principal component analysis (PCA) (Figure 1) suggests that gallic acid was the polyphenol most influenced by the elicitor treatments, amongst which the cold treatment had the greater weight. Cold tolerance is an intrinsic characteristic of *F.* × *ananassa*, a perennial plant, which is mostly grown in temperate climate zones, freezing in the winter and re-growing in spring [26]. Cold stress has been shown to induce polyphenol synthesis as antioxidants in strawberry fruits and other plants [27]. Given that the cold treatment induced large leaf size in vitro, which could translate to larger biomass, and considering that tropical highland strawberry crops rarely freeze, future experiments should consider the relevance of stimulating cold-regulated responses in the shikimate pathway for polyphenol accumulation, particularly using in vitro plant production systems.

Table 4. Morphological characterization (mean ± SD) of strawberry plants (*F.* × *ananassa* cv. Festival) in vitro (n = 30, except for control with n = 10) after treatment with three elicitors (UV irradiation, cold exposure, and cysteine) for 15 days.

Parameter	Control	UVC (254 nm)	Cold (4 °C)	Cysteine (200 mg L^{-1})
Length of the longest root (cm)	0.89 ± 0.93	1.49 ± 0.90	1.39 ± 0.48	1.33 ± 0.81
Length of the longest stem (cm)	2.14 ± 0.66	2.93 ± 0.99	2.66 ± 1.01	2.49 ± 0.89
Length of the longest leaf (cm)	0.94 ± 0.25	1.23 ± 0.26 *	1.21 ± 0.30 *	1.10 ± 0.32
Width of the longest leaf (cm)	0.99 ± 0.23	1.18 ± 0.29	1.06 ± 0.31	1.15 ± 0.33
Total amount of leaves	19.50 ± 5.10	38.53 ± 24.9 *	25.93 ± 9.22	21.30 ± 4.49
Total amount of buds	3.00 ± 0.82	5.47 ± 2.74 *	3.87 ± 1.76	4.47 ± 1.41
Fresh weight of leaves (g)	0.55 ± 0.28	0.59 ± 0.42	0.39 ± 0.37	0.36 ± 0.24

Differences ($p < 0.05$) in treatments relative to their respective control were analyzed by one-way ANOVA followed by Dunnett's test (GraphPad Prism v. 10.1.1., USA) and are marked by asterisks.

Table 5. Morphological characterization (mean ± SD) of strawberry (*F.* × *ananassa* cv. Festival) greenhouse-grown seedlings (n = 6) after treatment with three elicitors (UV irradiation, cold exposure, and foliar cysteine) for 15 days.

Parameter	Control	UVC (254 nm)	Cold (4 °C)	Cysteine (200 mg L^{-1})
Length of the longest root (cm)	11.30 ± 1.47	10.78 ± 1.63	14.08 ± 2.66 *	13.68 ± 0.85
Length of the longest stem (cm)	7.28 ± 0.54	8.96 ± 1.35	9.52 ± 0.39 *	9.07 ± 1.15 *
Length of the longest leaf (cm)	3.47 ± 0.37	3.20 ± 0.49	3.63 ± 0.16	3.50 ± 0.62
Width of the longest leaf (cm)	3.03 ± 0.20	2.78 ± 0.37	3.20 ± 0.20	2.95 ± 0.36
Total amount of leaves	12.67 ± 3.14	13.50 ± 1.64	14.33 ± 3.50	16.00 ± 4.10
Total amount of buds	4.33 ± 1.21	5.17 ± 0.75	5.33 ± 1.21	5.83 ± 2.04
Fresh weight of leaves (g)	1.08 ± 0.21	0.98 ± 0.23	1.37 ± 0.41	1.35 ± 0.59

Differences ($p < 0.05$) in treatments relative to their respective control were analyzed by one-way ANOVA followed by Dunnett's test (at 15 days) (GraphPad Prism v. 10.1.1., USA) and are marked by asterisks.

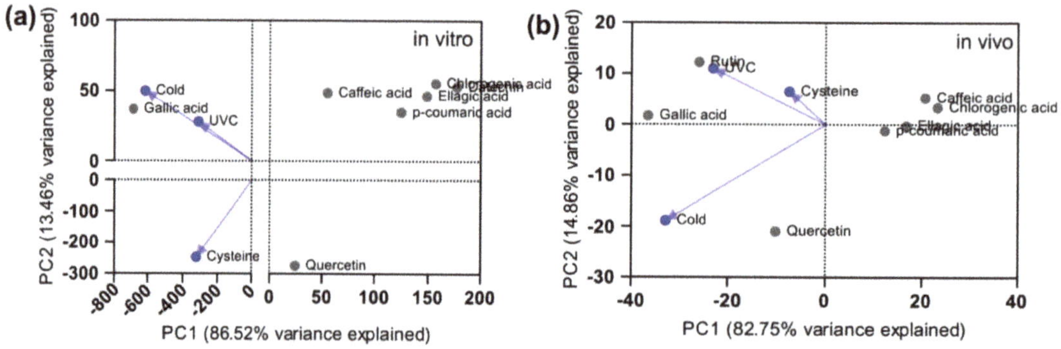

Figure 1. Principal component analysis (PCA) of specific polyphenol content in strawberry (*Fragaria* × *anannasa* cv. Festival) leaves from plants grown in vitro (**a**) and in vivo (**b**) after treatment with three elicitors (UV irradiation, cold exposure, and foliar cysteine) for 15 days (GraphPad Prism v. 10.1.1., USA).

To the best of our knowledge, this is the first report comparing the polyphenol profile of *F.* × *ananassa* cv. Festival leaves in plants grown in a tropical highland environment under three polyphenol-eliciting treatments in vitro vs. their counterparts in vivo, showing higher amounts and faster accumulation of polyphenols in the in vitro regenerated plants. This information should be considered in further experiments using other biotechnological strategies, such as large-scale cultivation in bioreactor systems.

3. Materials and Methods

Reagents and chemicals. The sodium hypochlorite (3%), citric acid (>99% purity), and ascorbic acid (99% purity) were purchased from LABQUIM CR (Alajuela, Costa Rica). The Gamba Oxi® (2% hydrogen peroxide, 1% peracetic acid, 2% acetic acid) was purchased from RACKAM S.A. (Alajuela, Costa Rica). The Agri-mycin® 16.5 WP (oxytetracycline + streptomycin sulfate) was from Pfizer S.A. (México-Toluca, México). The Afungil® 50 WP (benomyl) was from SERACSA (Cartago, Costa Rica). The 6-benzylaminopurine (99% purity) and Murashige & Skoog macro- and micronutrient salt bases were obtained from PhytoTech Labs (Lenexa, KS, USA). The L-cysteine, Folin & Ciocalteu's phenol reagent and gallic acid, (+)-catechin, caffeic acid, chlorogenic acid, *p*-coumaric acid, ellagic acid, quercetin, and rutin (all standards > 99% purity) were obtained from Sigma-Aldrich (St. Louis, MO, USA). The methanol and ethyl acetate (>99% HPLC-grade) were from Honeywell Riedel-de-HaënTM (Muskegon, MI, USA).

Plant material. The strawberry (*F.* × *ananassa* cv. Festival) plants were obtained from Fresas de Altura S.A. in Llano Grande, Cartago, Costa Rica (permit R-CM-ITCR-001-2022-OT-CONAGEBIO), they were kept in a greenhouse at the Costa Rica Institute of Technology (Cartago, Costa Rica) under natural temperature (17–25 °C) and light (~12 h daylight) conditions (August to December 2021), and they were watered daily in the morning and in the afternoon.

In vitro culture. The introduction and multiplication were based on the protocols by Jiménez-Bonilla & Abdelnour-Esquivel [28]. The explants introduced in vitro consisted of runners with apical meristems and runners with axillary meristems (approximately 3 cm long). For disinfection, explants (n = 24, each) were treated with surgical soap, water, Agri-mycin® + Afungil®, and a final step of sodium hypochlorite (treatment A) or Gamba Oxi (treatment B) (Table 6). After disinfection, the explants were rinsed three times with sterile distilled water in aseptic conditions and kept in an antioxidant solution of sterile citric acid (300 mg L^{-1}) and ascorbic acid (300 mg L^{-1}) while they were introduced in cylindrical test tubes containing 10 mL of a complete standard Murashige & Skoog (MS) medium [29] without growth regulators. The pH was adjusted to 5.7 with NaOH 1 M. The explants were incubated in a clean room for 8 days at 27 °C in darkness, followed by light exposure (with a photoperiod of 16 light hours at 2000 lux and 8 h in diffuse light) for 4 weeks. Contamination was evaluated every week. The aseptically established plants were subsequently grown in complete MS medium with 6-benzylaminopurine (BAP) in different concentrations (0.5, 1, and 1.5 mg L^{-1}) and subcultivated every 6 weeks.

Table 6. Disinfection protocol of strawberry explants for in vitro culture.

Treatment	Disinfectant Steps	Working Concentration	Incubation Time in Agitation (min)
A	Surgical soap	-	10
	Water	-	5
	Agri-mycin® + Afungil®	5 g L^{-1}, 4 g L^{-1}	15
	NaClO	3%	15
B	Surgical soap	-	10
	Water	-	5
	Agri-mycin® + Afungil®	0.8 g L^{-1}, 0.8 g L^{-1}	15
	Gamba Oxi	40%	5

Elicitor treatments. Three elicitor treatments (UV irradiation, cold exposure, and cysteine) were tested in strawberry (*F.* × *ananassa* cv. Festival) seedlings kept in a greenhouse or in in vitro plants. For the cold treatment, the plants were incubated at 4 °C for 12 h daily in the dark. The UV treatment consisted of irradiating the plants with 254 nm (UV-C) for 7 min each day at the same hour. Finally, 7 mL foliar cysteine (200 mg L^{-1}) was applied daily over the leaves of each seedling, or it was added once to the semisolid medium (in vitro plants) at the same concentration. These treatments were established from preliminary tests, based on practicality, local availability, and cost. The seedlings were exposed to each of the three elicitors for 15 days (in vivo and in vitro) to evaluate their individual effects. However, due to minor changes in plant morphology, a second experiment was conducted exposing the plants to all elicitors combined for 30 days (in vivo). The morphological features were measured in all plants at the end of the experiments.

Polyphenol extraction. All the leaves from the plants in vivo and in vitro were freeze-dried and pulverized in a food processor. Insoluble (cell-wall-bound) and soluble (free, soluble ester- and glycoside-bound) polyphenols were extracted using aqueous methanol and ethyl acetate, respectively, as described before [30,31]. In the in vitro plants, due to low biomass, the dried leaves had to be pooled together in three groups of 10 replicates each (n = 3), resulting in a single control sample; and the dried material was only solubilized in methanol (10 mg mL^{-1}) before the chromatographic analysis.

Total polyphenol content (TPC). TPC was determined in the water–ethanol extract (in vivo plants) or in the methanol-solubilized leaves (in vitro plants) by the Folin–Ciocalteu method [32]. TPC was calculated by comparing with an external calibration curve of gallic acid (10–80 mg L^{-1}, r^2 = 0.9863) (>99% purity, Sigma-Aldrich, St. Louis, MO, USA).

Chromatographic analysis. The methanol-dissolved samples were analyzed in an ultra-high-performance liquid chromatography system (UHPLC) with a triple quadrupole mass spectrometer (MS/MS) and a diode-array detector (DAD, Ultimate 3000, TSQ Endura, series TQH-E1-0288, Thermo Fisher Scientific, Waltham, MA, USA), as previously reported [31,33]. Mass detection of polyphenols was carried out with an electrospray ionization (ESI) source operating in negative polarity with SRM scan mode with a cycle time of one second. Mass settings were as follows: spray voltage, 2500 V (negative); sheath gas flow, 50 arb; auxiliary gas flow, 10 arb; sweep gas flow, 1 arb; ion transfer tube temperature, 300 °C; vaporizer temperature, 350 °C; and collision induced gas, 1.5 mTorr. Peaks were analyzed with the Chromeleon software (version 7.3, Thermofisher Scientific, Waltham, MA, USA) and identified and quantified by UV–vis wavelengths and mass spectra using the external curves of authentic polyphenol standards (Sigma-Aldrich, St. Louis, MO, USA). At least four samples were analyzed per treatment and each one was injected once for UHPLC-MS/MS-DAD analysis.

Statistical analysis. Normality was tested and the differences between the untreated controls and the elicitor treatments were calculated with a two-tailed unpaired t test, a one-way analysis of variance (ANOVA) or with a Kruskal–Wallis test, with the corresponding post hoc tests. A principal component analysis (center method; principal components selected based on eigenvalues greater than 1) was conducted to evaluate relationships between the elicitor treatments and the polyphenol content. All statistical analyses were performed with Graph-Pad Prism (v. 10.1.1., GraphPad Software, San Diego, CA, USA).

Author Contributions: L.A.C.-C. and K.S.-A. designed the study. K.S.-A. and M.F.B.-Z. conducted the elicitor experiments in vivo. K.S.-A., L.S.-C. and A.A.-E. conducted the elicitor experiments in vitro. K.S.-A. and A.I.-M. conducted the HPLC analysis. L.A.C.-C., K.S.-A. and F.V.-R. wrote the first draft, and all authors read, edited, and approved the final manuscript. F.V.-R. is responsible for the final content. All authors have read and agreed to the published version of the manuscript.

Funding: This project was funded by Vicerrectoría de Investigación y Extensión, Instituto Tecnológico de Costa Rica.

Institutional Review Board Statement: Not applicable.

Informed Consent Statement: This article does not contain any studies with human or animal subjects performed by any of the authors.

Data Availability Statement: Datasets are available from the corresponding author on reasonable request.

Acknowledgments: This We thank our students, M. J. Rodríguez-Brenes and N. Vílchez-Mejía, for their valuable technical assistance.

Conflicts of Interest: The authors declare no conflicts of interest.

eferences

Sharma, A.; Shahzad, B.; Rehman, A.; Bhardwaj, R.; Landi, M.; Zheng, B. Response of phenylpropanoid pathway and the role of polyphenols in plants under abiotic stress. *Molecules* **2019**, *24*, 2452. [CrossRef] [PubMed]

Rasouli, H.; Farzaei, M.H.; Khodarahmi, R. Polyphenols and their benefits: A review. *Int. J. Food Prop.* **2017**, *20*, 1700–1741. [CrossRef]

Cory, H.; Passarelli, S.; Szeto, J.; Tamez, M.; Mattei, J. The Role of Polyphenols in Human Health and Food Systems: A Mini-Review. *Front. Nutr.* **2018**, *5*, 87. [CrossRef] [PubMed]

Fraga, C.G.; Croft, K.D.; Kennedy, D.O.; Tomás-Barberán, F.A. The effects of polyphenols and other bioactives on human health. *Food Funct.* **2019**, *10*, 514–528. [CrossRef] [PubMed]

Bhaskar, R.; Xavier, L.S.E.; Udayakumaran, G.; Kumar, D.S.; Venkatesh, R.; Nagella, P. Biotic elicitors: A boon for the in-vitro production of plant secondary metabolites. *Plant Cell Tissue Organ. Cult.* **2022**, *149*, 7–24. [CrossRef]

Halder, M.; Sarkar, S.; Jha, S. Elicitation: A biotechnological tool for enhanced production of secondary metabolites in hairy root cultures. *Eng. Life Sci.* **2019**, *19*, 880–895. [CrossRef]

Hano, C.; Tungmunnithum, D. Plant Polyphenols, More than Just Simple Natural Antioxidants: Oxidative Stress, Aging and Age-Related Diseases. *Medicines* **2020**, *7*, 26. [CrossRef] [PubMed]

Qin, Z.; Liu, H.M.; Ma, Y.X.; Wang, X.D. Developments in Extraction, Purification, and Structural Elucidation of Proanthocyanidins (2000–2019). *Stud. Nat. Prod. Chem.* **2021**, *68*, 347–391.

Gündüz, K. Chapter 30 - Strawberry: Phytochemical Composition of Strawberry (*Fragaria* × *ananassa*). In *Nutritional Composition of Fruit Cultivars*; Simmonds, M.S.J., Preedy, V.R., Eds.; Academic Press: San Diego, CA, USA, 2016; pp. 733–752.

0. Fierascu, R.C.; Temocico, G.; Fierascu, I.; Ortan, A.; Babeanu, N.E. Fragaria genus: Chemical composition and biological activities. *Molecules* **2020**, *25*, 498. [CrossRef] [PubMed]

1. Giampieri, F.; Forbes-Hernandez, T.Y.; Gasparrini, M.; Alvarez-Suarez, J.M.; Afrin, S.; Bompadre, S.; Quiles, J.L.; Mezzetti, B.; Battino, M. Strawberry as a health promoter: An evidence based review. *Food Funct.* **2015**, *6*, 1386–1398. [CrossRef]

2. Afrin, S.; Gasparrini, M.; Forbes-Hernandez, T.Y.; Reboredo-Rodriguez, P.; Mezzetti, B.; Varela-López, A.; Giampieri, F.; Battino, M. Promising Health Benefits of the Strawberry: A Focus on Clinical Studies. *J. Agric. Food Chem.* **2016**, *64*, 4435–4449. [CrossRef] [PubMed]

3. Villamil-Galindo, E.; Van de Velde, F.; Piagentini, A.M. Extracts from strawberry by-products rich in phenolic compounds reduce the activity of apple polyphenol oxidase. *LWT Food Sci. Technol.* **2020**, *133*, 110097. [CrossRef]

4. Villamil-Galindo, E.; Van de Velde, F.; Piagentini, A.M. Strawberry agro-industrial by-products as a source of bioactive compounds: Effect of cultivar on the phenolic profile and the antioxidant capacity. *Bioresour. Bioprocess.* **2021**, *8*, 61. [CrossRef] [PubMed]

5. Salas-Arias, K.; Irías-Mata, A.; Sánchez-Kopper, A.; Hernández-Moncada, R.; Salas-Morgan, B.; Villalta-Romero, F.; Calvo-Castro, L.A. Strawberry *Fragaria* × *ananassa* cv. Festival: A Polyphenol-Based Phytochemical Characterization in Fruit and Leaf Extracts. *Molecules* **2023**, *28*, 1865. [CrossRef] [PubMed]

6. Calvo-Castro, L.A.; Lobo-Vázquez, M.; Gómez-González, J.C.; Arnáez-Serrano, E.; Zamora-Fallas, G.; Sánchez-Zúñiga, K.; Centeno-Cerdas, C. Bioactive potential of tropical highland apple (*Malus domestica* cv. Anna) crude extract: Opportunities for food waste revalorization. *Future J. Pharm. Sci.* **2022**, *8*, 57. [CrossRef]

17. Li, H.; Tsao, R.; Deng, Z. Factors affecting the antioxidant potential and health benefits of plant foods. *Can. J. Plant Sci.* **2012**, *92*, 1101–1111. [CrossRef]

18. Wilson, S.A.; Roberts, S.C. Recent advances towards development and commercialization of plant cell culture processes for the synthesis of biomolecules. *Plant Biotechnol. J.* **2012**, *10*, 249–268. [CrossRef] [PubMed]

19. Naik, P.M.; Al-Khayri, J.M. Impact of Abiotic Elicitors on In vitro Production of Plant Secondary Metabolites: A Review. *J. Adv. Res. Biotechnol.* **2016**, *1*, 1–7.

20. Marchev, A.S.; Yordanova, Z.P.; Georgiev, M.I. Green (cell) factories for advanced production of plant secondary metabolites. *Crit. Rev. Biotechnol.* **2020**, *40*, 443–458. [CrossRef] [PubMed]

21. Garza-Alonso, C.A.; Olivares-Sáenz, E.; González-Morales, S.; Cabrera-De la Fuente, M.; Juárez-Maldonado, A.; González-Fuentes, J.A.; Tortella, G.; Valdés-Caballero, M.V.; Benavides-Mendoza, A. Strawberry Biostimulation: From Mechanisms of Action to Plant Growth and Fruit Quality. *Plants* **2022**, *11*, 3463. [CrossRef] [PubMed]

22. Khan, A.K.; Kousar, S.; Tungmunnithum, D.; Hano, C.; Abbasi, B.H.; Anjum, S. Nano-elicitation as an effective and emerging strategy for in vitro production of industrially important flavonoids. *Appl. Sci.* **2021**, *11*, 1694. [CrossRef]

23. Guru, A.; Dwivedi, P.; Kaur, P.; Pandey, D.K. Exploring the role of elicitors in enhancing medicinal values of plants under in vitro condition. *S. Afr. J. Bot.* **2022**, *149*, 1029–1043. [CrossRef]

24. Matkowski, A. Plant in vitro culture for the production of antioxidants—A review. *Biotechnol. Adv.* **2008**, *26*, 548–560. [CrossRef] [PubMed]

25. Kotsupiy, O.; Karpova, E.; Trofimova, E.; Novikova, T.; Ambros, E. Transformation of Strawberry Plants' Phenolic Profile after Treatment with a Mechanocomposite Based on Silicon Chelates in the Course of Development under In Vitro, Ex Vitro, and In Vivo Conditions. *Horticulturae* **2023**, *9*, 157. [CrossRef]

26. Koehler, G.; Rohloff, J.; Wilson, R.C.; Kopka, J.; Erban, A.; Winge, P.; Bones, A.M.; Davik, J.; Alsheikh, M.K.; Randall, S.K. Integrative "omic" analysis reveals distinctive cold responses in leaves and roots of strawberry, *Fragaria* × *ananassa* 'Korona'. *Front. Plant Sci.* **2015**, *6*, 1–21. [CrossRef] [PubMed]

27. Koyama, R.; Ishibashi, M.; Fukuda, I.; Okino, A.; Osawa, R.; Uno, Y. Pre- and Post-Harvest Conditions Affect Polyphenol Conte in Strawberry (*Fragaria* × *ananassa*). *Plants* **2022**, *11*, 2220. [CrossRef] [PubMed]
28. Jiménez-Bonilla, V.; Abdelnour-Esquivel, A. Protocolo de micropropagación de arándano nativo de Costa Rica (*Vacciniu consanguinium*). *Rev. Tecnol. Marcha* **2018**, *31*, 144–159. [CrossRef]
29. Murashige, T.; Skoog, F. A revised medium for rapid growth and bioassay with tobacco tissue cultures. *Physiol. Plant* **1962**, *1* 474–497. [CrossRef]
30. Santiago, R.; Reid, L.M.; Arnason, J.T.; Zhu, X.Y.; Martinez, N.; Malvar, R.A. Phenolics in maize genotypes differing in susceptibilit to *Gibberella* stalk rot (*Fusarium graminearum* Schwabe). *J. Agric. Food Chem.* **2007**, *55*, 5186–5193. [CrossRef]
31. Lux, P.E.; Freiling, M.; Stuetz, W.; Von Tucher, S.; Carle, R.; Steingass, C.B.; Frank, J. (Poly)phenols, Carotenoids, and Tocochro manols in Corn (*Zea mays* L.) Kernels As Affected by Phosphate Fertilization and Sowing Time. *J. Agric. Food Chem.* **2020**, *68* 612–622. [CrossRef]
32. Rojas-Garbanzo, C.; Pérez, A.M.; Vaillant, F.; Pineda-Castro, M.L. Physicochemical and Antioxidant Composition of Fresh Peac Palm (*Bactris gasipaes* Kunth) Fruits in Costa Rica. *Braz. J. Food Technol.* **2016**, *19*, e2015097. [CrossRef]
33. Fratianni, F.; Cardinale, F.; Cozzolino, A.; Granese, T.; Albanese, D.; Di Matteo, M.; Zaccardelli, M.; Coppola, R.; Nazzaro, Polyphenol composition and antioxidant activity of different grass pea (*Lathyrus sativus*), lentils (*Lens culinaris*), and chickpea (*Cicer arietinum*) ecotypes of the Campania region (Southern Italy). *J. Funct. Foods* **2014**, *7*, 551–557. [CrossRef]

MDPI AG
Grosspeteranlage 5
4052 Basel
Switzerland
Tel.: +41 61 683 77 34

Molecules Editorial Office
E-mail: molecules@mdpi.com
www.mdpi.com/journal/molecules

www.ingramcontent.com/pod-product-compliance
Lightning Source LLC
La Vergne TN
LVHW070445061126
851858LV00011B/166